1995

Television Careers: A Guide to Breaking and Entering

by Linda Guess Farris

BUY THE BOOK
ENTERPRISES

Television Careers: A Guide to Breaking and Entering

by Linda Guess Farris

Published by:
Buy the Book Enterprises
182 Canyon Road
Fairfax, CA 94930

Cover: Rob Grant
Interior Design: Marisa Carder
Editor: Helyn Pultz

First Printing 1995
Printed and bound in the United States of America

Library of Congress Cataloging-in-Publication Data
Farris, Linda Guess
Television Careers: A Guide to Breaking and Entering / by Linda Guess Farris
p. cm.
Includes index.
1. Television—Vocational guidance—United States. I. Farris, Linda Guess
II. Title
LCCC# 95-075581
ISBN 0-9638673-1-8: $14.95 Softcover

Dedication

TO DENNIS FITCH

I owe my career in television to one person. Dennis Fitch hired me, not once, but twice. He gave me my first job in TV and then my second. Those two jobs spanned 18 years. Since 1987, he's been president of his own media consulting firm in L.A. Much love to my lifelong friend and mentor, and to Marge for sharing him with the rest of us.

Contents

Preface

Television is the primary source of news for most Americans, the primary entertainment, the babysitter, the companion and the window on the world. It puts us smack dab in the center of a raging fire, on the front line of a war zone, in a helicopter over L.A. freeways watching the real-life drama of O.J.'s last run, and it gives us the best seat in the house at concerts, sporting events and courtrooms. And if real life isn't exciting enough, it gives us the soaps.

The only thing more exciting than having instant access to so many real and imaginary worlds through our remote control is being a part of the business that delivers those worlds. The TV environment thrives on pressure, deadlines, creativity and responsibility. It's a center of power, influence, money and celebrity. In fact, a 1986 poll of 1550 folks reports that Americans get more pleasure from television than from anything else—including sex, food, marriage, hobbies, vacations, sports or money!

It's little wonder that thousands of Americans want to join its ranks each year. And if you're one of those thousands, *Television Careers: A Guide to Breaking and Entering* will help you decide if a television career is really right for you. It offers valuable advice from those who have succeeded as well as those upon whom your success depends.

As public relations director at KRON-TV, San Francisco's NBC affiliate, I received several calls a week from individuals desperately seeking information. They didn't necessarily want jobs in my area—TV promotion and publicity. Most wanted to work in news or programming, some in sales or engineering.

After many years of giving informational interviews, I hatched a clever plan. Once a month, I would host a free two-hour career panel and invite station employees to speak to anyone who wanted to listen. A handful of us "regulars" from news, creative services, public affairs and programming stayed after work to talk about the kinds of jobs that exist, offer hints for landing interviews, tell the stories of how we got in and answer questions. A representative from the station's Human Resources department used the forum to disseminate information about the station's job line, the procedure for job application, internships and the minority trainee program.

We began each career panel with an 18-minute video entitled, "How We Do the News." This fast-paced video featured reporter Vic Lee and news photographer Gary Mercer screeching out of the garage on their way to a demonstration, arriving late and trying to salvage the story; anchor Sylvia Chase dashing to the news set with a full five seconds to spare; and Fast Freddy Bushardt directing "the 5" (o'clock newscast) from the control room in what he refers to as a half-hour of "controlled mayhem." The video provided an excellent idea of life in a bustling newsroom and helped aspiring journalists decide if this was what they had in mind. Although the career panels gave a great deal of attention to the news department, since that was "home" to almost half the station's staff, those of us in other departments explained that there were lots of other opportunities for those who preferred a slightly less frantic existence.

Some job seekers returned month after month, hearing the same advice from us regulars, just to establish relationships with people on the inside. After two years of career panels, Nick Condos finally landed a job in office services (the proverbial mailroom). After one year, Hael Kobayashi was hired as a news production assistant on weekends. A couple of others found themselves in the right place at the right time and got their break after only one or two visits.

As the organizer of the career panels, I picked up a great deal of information over the years. I was invited to speak at universities and colleges about breaking into the business and even appeared on the local CBS affiliate's talk show, "People Are Talking," to pass on some of what I'd learned.

Now I've decided to put everything I've learned over my 20-year TV career into this book. And I've done what all TV job seekers should do. I've talked to people from small markets to large all over the country to draw on their experience and add their advice for today's aspiring TV journalist, program producer, sales manager, technician, publicist or graphic designer. You can think of this book as several dozen informational interviews rolled into one. Whether it's the beginning or the next step in your search, it will help you on your path.

Happy job hunting!

Acknowledgments

I owe a great debt of gratitude to so many for their contributions to this work. First, I thank those who allowed me to borrow their expertise and use their words. I couldn't have done it without them. I've listed them in alphabetical order with their titles and companies so that as you read the book, you can refer back to their affiliations. I particularly want to acknowledge Professor John Hewitt, Rosemary Wesela, Joe Fragola and Bill Groody whom I quote extensively.

I'm also indebted to those who read the manuscript and offered their comments and edits: Professors John Hewitt, George Mangan, Dr. Richard Taylor and Dr. Steve Guisinger, my terrific cousin. Mega-thanks to my sister and biggest fan, Kathy Clinton, a dern good proofer in her own right, and three of my best publicist hires of all time—Jodie Chase, Shari Jackman and Nancy Neustadt Goldstein. If you only knew how many booboos they caught! Thanks, too, to other critics/proofers/supporters—Darryl Compton, Mariko Todd, Carol Blackman, Marisa Carder, Jeff Marcus, Jeanmarie Murphy, Pat Giammarise, Susan Sikora, Fred Zehnder, Anna Chavez, Lisa Heft, and my long-time friend and colleague, Javier Valencia. And especially to the pro, Helyn Pultz, who was once my editor at *TV Guide,* and is now a book editor in Cambridge, Massachusetts. Finally, another special thanks to

Shirley Davalos, best bud, media coach, famous producer and moral supporter.

CONTRIBUTORS

Wendy Burch, Weekend Anchor/Reporter, WXIX-TV, Cincinnati
Rosy Chu, Director of Community Relations and Public Service, KTVU, Oakland
Darryl Compton, Director of Operations, CKS Pictures, Cupertino, CA
Lynn Costa, Director of Organization Management and Resources Planning, NBC, New York
Shirley Davalos, Executive Producer, Orion Express, Sausalito, CA
Joseph DiSante, Manager of West Coast Administration, ABC Television, Los Angeles
Paul A. Dunn, Director of Corporate Communications/Planning, WCNY, Syracuse
Don Fitzpatrick, President, Don Fitzpatrick Associates, San Francisco
Joe Fragola, Executive Producer, BayTV, San Francisco
Duane Fulk, freelance video editor
Jim Gaughran, News Producer, KRON, San Francisco
Mike Gaynes, Newswriter, Producer, KRON and KTVU
Janette Gitler, Director of Local Programming and Community Relations, KRON
Vernon Glenn, Sports Anchor, KRON
Bill Groody, President, North Country Communications, Lakeport, CA
Deborah Guardian, Assistant to the Station Manager, KQED, San Francisco
Lisa Heft, Community Education Consultant
John Hewitt, Professor, Broadcast and Electronic Communications Arts, San Francisco State University
Elizabeth Upham Howell, Publicity Coordinator, CNN, Atlanta
Betty Hudson, Executive Producer, NBC Productions, New York
Shari Jackman, former Publicist, KRON

John James, Program Manager, Syracuse NewChannels
Glen Kinion, Photographer/Editor, KRON
Barrett Lester, Program Manager, Continental Cablevision, Lawrence, MA
Douglas McKnight, News Director, KCCN, Monterey, CA
Libby Moore, Personal Assistant to Maury Povich, New York
Diana Mordock, Small Business Marketing Consultant
Daniel Murphy, Engineer, KRON
Elena Nachmanoff, Vice President, Talent Development for NBC News
Soledad O'Brien, Reporter, KRON
Jim Owens, Director of Engineering, WLS, Chicago
Aimee Rosewall, former News Assignment Editor, KRON; currently works in multimedia
Michael Scannell, Audience Service Representative, WGBH, Boston
Lisa Sanchez-Corea Simpson, Director of Business Development and Multimedia, NBC News, New York
Jacqueline Trube, Internship Coordinator, Turner Broadcasting System, Inc., Atlanta
Ruby Petersen Unger, Educational Filmmaker
Suzy Polse Unger, Vice President of Development, Buena Vista Productions, Burbank
Javier Valencia, Community Relations Manager, KRON
Rosemary Wesela, Executive Secretary to Vice President, News, KRON
Evan White, Senior Anchor, BayTV, San Francisco
Andrea Wishom, Audience Supervisor, Harpo Productions, Chicago
Fred Zehnder, News Director, KTVU

Casing the Joint

How Local Television Stations Operate

WARNING: Read these important instructions before operating this book!

I know you're probably champing at the bit to get on with it. You want to know *right now* how to get that high-paying anchor job, or at least that internship. Skip ahead, if you must. But I'm warning you: Reading this background could make the difference between your success and failure.

When I first decided I wanted to work in TV, I had no clue how the business operated. And I committed the ultimate faux pas: I ventured out on informational interviews unprepared. My lack of basic knowledge made it fairly obvious that I had no idea where I might fit in and what I might have to contribute.

Informational interviews are hard to get. To take advantage of the ones you do land, you must do your homework. By being prepared, you make the most valuable use of your time and your interviewer's time. You present yourself in a favorable light, being informed and knowledgeable. You ask intelligent questions and your interviewer is much more likely to think you'll make a valuable addition to his or her team.

Doing your homework means finding out how the business works and learning the language (see chapter 14). So, how do local commercial stations operate?

Trick question... Every single one operates differently. One of the primary reasons is that TV stations vary in size from just a handful of employees in the smallest markets to around 250 or 300 in the largest markets. In the early '80s, a few local stations were inching towards the 400 mark, but that's before the advertising revenue pie was being shared with cable stations and dozens of other new advertising outlets. With the increased competition came the belt tightening of the '90s and the euphemism of "downsizing," a much more socially acceptable term than head-rolling or axing. The large market stations are now combining responsibilities and departments, and the gap in staff size has shrunk.

In a few of the largest markets, the ABC, NBC and CBS stations are owned and operated by the network. They're often simply called O&Os. All other local commercial stations are affiliates or independents. Affiliates are the distribution channels of the networks. The networks provide programming to their stations at no charge. In fact, they even pay a fee to their local stations for carrying their programming. Networks, however, feed more than programming. They also feed commercials adjacent to their programs. The revenue from that nationally-sold commercial time supports the networks.

The "big three" networks fill a little more than a third of their affiliates' programming day, between 8 and 10 hours. (Fox provides slightly less.) Weekday network programming consists of a two-hour morning news show from 7 to 9 and the national nightly news, soap operas, prime time fare and late night entertainment such as "Late Show with David Letterman" and "The Tonight Show." Add to that, of course, network coverage of sporting events and major news stories, which preempts whatever is in its path.

Local stations fill another several hours of programming with syndicated product supplied by independent production companies. These include most of the talk show fare: "Oprah," "Donahue," "Sally Jessy Raphael," "Live! with Regis & Kathie Lee," "Geraldo," "Maury Povich," "Ricki Lake"; many game shows such as the long-running "Jeopardy!" and "Wheel of Fortune"; "infotainment" programs including "Entertainment Tonight," "Siskel & Ebert," "Hard Copy," "A Current Affair"; reruns of sitcoms, "Star Trek" and old movies.

Independent stations, those with no network affiliation, rely primarily on syndicated programming, although some are now affiliating with budding networks, including the United Paramount Network and Warner Bros. Network.

Very few employees are needed at the local level to broadcast the network and syndicated programming. Both network and syndicated product are fed to the station by satellite, and either relayed instantaneously to the transmitter and then to the viewer's home via the air or cable, or taped for later transmission. This requires only pressing a button or two by the technician or engineer in charge.

So what's left for the local commercial station?

In a word, *news*. Okay, in two words: news and information. In the '60s and '70s, local TV stations were producing much more local non-news programming. Most stations in mid-size to large markets had a local daily talk show, weekly public affairs series, several had their own children's programming and a few had local game shows. Much of this programming disappeared in the '80s. Some programming departments have merged into news departments and the occasional documentary or local special will fall under the overall responsibility of the news director. You still find some public affairs programming at the local level, occasionally in the form of a weekly local issues program, often hosted by news anchors or reporters, or by the public affairs director.

As the viewing public has been wowed by the glitz of fast-paced TV with high production values funded at the national level, local programming has had a tough time competing. It's much more cost-effective for individual stations to purchase the rights to a program such as "Hard Copy," and in essence, share the cost with stations in dozens of other markets where the show is sold, than to produce it locally just for that station's air. And what the final product sacrifices in local orientation, it makes up for in slick production values, big name hosts or celebrities, or travel afforded by the larger overall budget of the show. Case in point: a local station could never match the production budget of, say, "Lifestyles of the Rich and Famous." The proliferation of syndicated talk and infotainment shows has made it more difficult for local station talk show producers to attract celebrities. Talk show guests are seen by a much larger audience if they appear on nationally syndicated shows aired in markets all over the country, than if they travel from city to city to appear on individual shows. Promotional tours are also expensive and time-consuming. (If you're interested in working for a nationally syndicated show, check out chapter 26.)

Noncommercial stations, such as local Public Broadcasting System (PBS) affiliates, differ in two dramatic ways from local commercial stations. Few PBS affiliates have local news operations. Also, they don't rely on commercial advertising, so they aren't focused on the profitability of specific programming. On the other hand, PBS stations are concerned with raising funds, through membership drives and grants, to support the educational, cultural and entertainment programming they provide.

Local cable operators, like PBS affiliates, rarely have local news operations. Many do have Local Origination (programming) departments. Cable is a fast-growing television industry with all kinds of career opportunities. Chapters 23 and 24 are devoted to cable operators and cable networks.

NEWS

In a commercial TV station, news is the most visible department. Not only is it the largest, often comprising almost half of a station's employees, but many of the other departments—operations, engineering, creative services or promotion, the graphics or design department—work to support or promote the news product. The news department is usually home to the most recognizable station employees—the news reporters and anchors—some of whom make more money than their bosses. The nature of news, its mission to provide the public with important information in a timely fashion, the urgency of breaking stories, of exposing corruption, being our eyes and ears on the world, gives it its stature.

Local news has remained the monopoly of local TV since it's one area that outside production companies cannot usurp. You've got to be in town to cover the town's news. News also draws higher ratings than most, if not all, other locally produced programming and is the source of much of the station's advertising revenue. Yes, it costs a lot to pay for the cameras, the remote vans, the salaries, the computers, but look at it this way. You use the same sets day after day, you can often repackage the video for more than one newscast, your staff is available to you around the clock and you're filling a few hours of programming a day. So despite the cost, it's still very profitable.

Another benefit of local news is that it gives a station its local identity through its anchors and reporters. Stations are always struggling for viewer loyalty and they want anchors with high "Q" ratings (recognizability quotients) to whom the viewers return night after night.

So how does the news department operate?

This description applies to almost all TV news departments. The primary difference is the number of people doing each job. Whereas a top ten market station in, say, Atlanta or Boston, might have 15 news producers on staff, a station in Lafayette, Indiana, the 190th market, might have one, who also quadruples as the writer, shooter and news director.

Local TV news gathering is a 24-hours-a-day proposition. Large market newsrooms are staffed around the clock. In the past decade or two, many have added early morning newscasts before the network morning shows and/or 24-hour news updates, both requiring staffs. In smaller markets, the staff may not actually be at the station at night, but they are never far from their beepers or phones.

The news operation is headed by the news director, who, along with the managing editor, is responsible for the overall editorial content of the newscasts. The newscasts consist of local news combined with stories from national news wires and network news services.

Getting the story

The central nervous system of the newsroom is the assignment desk, headed by an assignment manager or the managing editor. The desk staff consists of assignment editors, and depending on the size of the news operation, desk assistants, a news planner and interns. The desk is manned or womanned around the clock by an assignment editor, who is the receptor of all breaking news tips. The desk staff monitors police, fire and Coast Guard radio scanners. The desk assigns (hence the name *assignment desk*) reporters and shooters (news photographers) to cover breaking news events. In some cases, they just send a shooter on the story to get video which will later be explained by anchor-read copy. In a smaller newsroom, the reporter and shooter may be one and the same, in which case he or she will set up the camera and then stand in front of it.

The assignment editor also selects individuals or teams to cover "scheduled" news events, such as parades, news conferences, demonstrations, public hearings, court hearings or the arrival of a winning sports team, major politico or celebrity.

The news planner, who's one of the assignment desk staff, sorts through the stacks of mail and faxes received daily from groups, businesses or individuals seeking coverage of their "news" or event. Due to time limitations and the sheer quantity of mail and faxes, much of it ends up in the circular file. To make the cut, the proposed news item needs to affect a certain percentage of the audience or be of interest to them. News managers know that to hang onto their jobs, they have to hold their audiences, i.e., worry about ratings. However, some purists (we call them journalists with a capital J) have a tough time covering fluff pieces, sensational stories that have been beaten into the ground or items that are added to the newscast to tie in with entertainment programming (such as "Stay tuned for an exclusive interview with the star of tonight's made-for-TV movie.") I remember one of our reporters who was livid that she was assigned to cover the "Wheel of Fortune" contestant search I was coordinating in a local hotel ballroom.

The assignment editors and desk assistants keep in touch with the crews in the field via pagers and cellular phones, and may reassign them to a more important breaking story while they're in the field. Those working the desk are always on the phone, checking facts, locations, details, taking calls from agencies, individuals and police departments concerning news tips.

Producing the newscast

Producers of the various newscasts determine, with input from news management, the content of their shows—which story will lead the newscast, how much time to allow for each story, whether it is a straight read, a live stand-up with video, or a package, and who will introduce which stories. The producer is responsible for timing the newscast to fit precisely within the allotted time. (In large stations, a producer usually produces only one of the daily newscasts. In smaller stations, s/he might produce more than one, and in the smallest markets may even serve as a reporter or anchor as well.)

Much of the news is written by newswriters, who do not appear on camera. They rewrite stories which come off the wire services, and they write the introductions (intros) to other stories fed from the network news services.

Reporting the story

Reporters must also be excellent writers, as they write the stories they're assigned to cover, and, when they're reporting "live," they need to be able to tell the story in a clear, concise way, without the "uhs" and "you-knows" of everyday conversation. It's much more difficult than it appears.

Depending on the station, anchors may do some reporting as well. They often host specials for the station, make public appearances and are involved in the community. Anchors usually work at least an eight-hour day even though you might see them on the air for just a half-hour or hour. They have to be familiar with all the news, because when a news video jams in the tape machine or they're covering a major breaking story, they have to be prepared to wing it. The anchor is often caught without a script and has to be able to think on his or her feet, and not only sound intelligent but actually tell the truth.

Getting the pictures

Shooters capture the stories on video. They are sent out either with a reporter or alone to get video on a story, news conference or event that doesn't require a reporter at the scene.

Video editors take several minutes of video from a story and piece together the pictures that tell the story *visually* in just around a minute. They work with the reporter or writer to deliver the edited piece.

Other technical people, who may actually be part of the engineering staff, are involved in the newscast itself. A director orchestrates getting the show on the air from the control room, giving direction to the technical director (TD), also called the switcher, who is responsible for rolling video, going live to a camera in the field, inserting graphics and superimposed identifiers (supers), and switching between cameras in the studios. The audio engineer handles all the sound, and sometimes an assistant director sets up live shots in the field and serves as the liaison between the crew in the field and the director. Others in the control room, or connected to it via headsets, operate the video playback machines, the character generator (which creates the supers), computer graphics, and manned or remote-control studio cameras.

Big news departments have a researcher or two, who can, in a matter of moments, locate file footage—video from the past which has all been encoded and cross-referenced in the computer. File footage is used for a related story or an update, or for obit stories. (When rock promoter Bill Graham died in a helicopter crash, San Francisco stations led their news with a retrospective of his life pulling together lots of file footage in short order.) The news resource role at TV stations is changing as computers give access to online information at everyone's fingertips.

Weather and sports, and sometimes investigative units, are subdepartments of news. In a large station, they might consist of a few employees each. If a station has a sports contract to cover local baseball games, the sports unit will be larger.

At the bottom of the pecking order in news is the production assistant or news assistant, and the intern. These are the gofers who rip scripts (separate the multiform printout of the script and collate it) for the director, the producer, the anchors and the floor manager; run late script changes to the control room; make phone calls; check facts; run errands; and, if they're smart, cheerfully do whatever is asked of them and more.

Read more about these positions in chapter 2, Entry Level Jobs, and chapter 6, Internships.

"There's more to life than news, weather and sports." Or so said a promotional slogan of San Francisco's KGO-TV in the '70s, although many of those whose lives are consumed by their news jobs don't really believe it. In fact, many seem oblivious that there are other departments in the station.

If you aren't an adrenaline addict in need of your daily fix of challenges, deadlines, and chaos that a newsroom provides, if you like a little more order in your life, you may find other departments more to your liking. Here are some of your other options in local television.

ENGINEERING/OPERATIONS

The engineering/operations department usually comprises about 30% of a station's employees. The department is headed by a chief engineer, and consists of managers who oversee the scheduling of the technical staff and station's studios and production facilities. The bulk of the department is made up of the technicians who actually operate and maintain the equipment which produces and transmits the news and programming through the airwaves to the tower and, from there, to your home, satellite dish or cable operator.

The technicians operate the cameras, audio boards, video tape equipment, robotic cameras, character generator and the video control board (switcher). They also serve as camera control and master control operators, directors, stage and floor managers, video editors and shooters. A large station might have as many as four or five maintenance engineers, who repair the equipment.

In the largest markets, these employees belong to a union such as the International Brotherhood of Electrical Workers (IBEW), the National Association of Broadcast Employees and Technicians (NABET) or the International Alliance of Theatrical Stage Employees and Motion Picture Machine Operators (IATSE).

Depending on station size and policy, the technicians might perform several different tasks, or even rotate among jobs. Although it's always been true in small markets, now news photographers in the large markets are having to learn to edit. Much of the specialization of the past is giving way to hybrid jobs that involve multiple skills.

PROGRAMMING

A big trend in the late '80s and '90s has been the disappearance of many stations' programming departments. In some cases, local programming has been replaced by more cost-effective syndicated programming. In other cases, local programming has been taken over by the station's news department. But yes, some programming departments are still thriving.

A program director oversees any local programming the

station produces, in addition to purchasing and scheduling syndicated programming. Even if the programming department no longer exists, someone must fulfill these purchasing and scheduling functions. Sometimes the general manager, station manager or news director takes on the important role of determining the program schedule.

Some stations have an executive producer to whom the various program producers report. The staff of a local program usually includes the producer, an associate producer/writer/researcher, a production assistant or two, and an intern or two. Although the technical crew (consisting of a shooter or two, a video editor, and, if it's a studio show, a director and floor manager) may work exclusively on one show, they usually report to the chief engineer.

Although most PBS stations don't produce local news, many of the larger ones do have local programming departments. Some also have educational divisions which provide programming to local school systems. For more on local programming produced by local cable operators, see chapter 23.

SALES

In a commercial television station, the sales department serves the vital function of selling commercial time, thus supplying the revenue that supports the entire operation—the news operation, the cost of syndicated programming, state-of-the-art equipment, the production of local programming, salaries, advertising and promotional campaigns. This includes delivering a healthy profit to the shareholders/owners.

PBS stations rely on fund development divisions to seek grant and underwriting revenue, and membership departments to raise the necessary operational funds through membership dues and viewer pledge drives. Local cable operators rely on a combination of commercial sales and viewer subscription fees for their revenue.

A commercial station's sales staff in a large market may consist of as many as 20 managers, account executives and sales assistants. The department is headed by a general sales manager or director of sales and is divided into national sales and local sales. National firms, such as Procter & Gamble, American Express and Coca-Cola, often buy time on networks as it's more cost-efficient than buying commercial time market by market. However, network show ratings (defined in chapter 14) aren't consistent from market to market. Often advertisers want to target specific markets, so they'll buy local stations to supplement their national buy. To assist with national sales, TV stations hire and work with national media rep firms such as Petry or Telerep. Approximately half of a TV station's sales are national sales. The other half is local sales, placed by media buyers from local advertising agencies or directly by local retailers who don't utilize an advertising agency.

The sales manager's roles include hiring and training the staff to sell those commercial spots; motivating, directing and setting objectives for staffers to reach; and, most importantly, forecasting business and determining rates. This means anticipating the market's need in the coming months, estimating what the volume of business will be.

The cost of TV time is totally demand-driven. The TV sales game is an excellent example of a supply and demand free market economy. A sales manager wants to set rates as high as the market will bear and still be able to sell all the inventory. If time passes and the available inventory isn't sold, the station has lost money. The time is irretrievable. If the rates are set too low, the time will be sold but the station will realize less revenue than if rates had been higher. A sales manager must anticipate the demand to maximize profit. The supply of available TV commercial time is limited to about 121/2 minutes per hour during local programming and much less during network programming when most of the spots are fed by the network. When demand is high, a newspaper can add pages. Not so in television. Because supply is limited, local commercial spot rates rise dramatically when there's high demand.

Rates change on a daily basis depending on inventory, how much time there is to sell and how much demand there is. Another factor in determining commercial rates is show ratings, how large an audience the station can deliver to the advertiser. (Ratings are explained in greater detail in the Sales Research section which immediately follows this one and in chapter 14, Talk the Talk.)

Account executives, called AEs for short, actually sell the time. They call on and service advertising agencies and local retail businesses, and in smaller markets they may get involved in producing the spots, helping coordinate the shoots and even writing scripts. Some TV stations use their own production and technical staffs to produce commercials for clients. AEs negotiate rates with their advertisers based on that client's volume of business and how much inventory the station has to sell relative to the overall demand. After "selling" the client, the AE then has to "sell" his or her sales manager on the deal made with the client. It's a high-stress, high-pressure job. In sales, you can't "BS" your success. It's measured on a daily basis in dollars and cents.

The sales assistants process orders and help write proposals, often working for two or more AEs or managers. The work is detail-oriented, high-end clerical and very important, as it's closely tied to the station's revenue.

In addition to spot sales, the station may sell sponsorships in public service campaigns, such as "For Kids' Sake" or "World of Difference," or in civic pride campaigns, such as "Spirit of Texas." These specially packaged sales vehicles usually encompass programming, public service messages, and often community outreach events. The station also may sell sponsorships of station-sponsored events or competitions, such as "Beat the Pro," a closest-to-the-pin golf competition. In return, the sponsors receive TV billboard mentions (still, full-screen graphics) adjacent to programs, signage at events and logos on public service announcements (PSAs) or promotional spots.

Occasionally, large TV stations are able to syndicate their own products or programs to other TV stations or cable networks.

In this way, they are acting in a fashion similar to a syndicator. With increased competition for advertising dollars, management at TV stations is constantly looking for new sources of revenue.

As mentioned earlier, a small percentage of a local network affiliate's revenue (perhaps 3-4%) is the fee it's paid by the network to air the network's programs and commercials.

SALES RESEARCH

Sales research is a small department, which, in even the largest markets, consists of only two or three people. Sometimes it's a subdepartment of sales. These researchers assist the account executives by keeping track of and interpreting the ratings, utilizing demographic information collected by the A.C. Nielsen Company, an independent ratings service, to analyze the audience for any given program. They help the AEs position the station's programming in the most favorable light. Brian Fiori, one of my all-time favorite research directors with an outrageous sense of humor, had a sign posted in his office at KRON: "Ratings distortion is no joke...it's my job."

TRAFFIC

The traffic department, which may also be a subdepartment of sales, publishes a daily program log which accounts for every second the station is on the air. It ranges in size from a couple to a dozen employees. A station which operates 24-hours-a-day can expect one day's log to list some 900 elements including programming, commercials, news updates, promotional spots, PSAs and station IDs. The programming and news blocks are logged first, then the remainder of the elements are provided by sales (sold commercial time), public affairs (public service announcements) and the promotion department (promotional spots). The computerized log is delivered to the master control room, the nerve center from which the station's signal is broadcast, to serve as the "bible" for the next day's air. Like sales, the traffic department's functions are tied closely to the station's revenue: The commercials must be correctly logged and aired without any glitches (problems with sound or image, the spot being cut off at the end) in order for the station to bill the advertiser for the time.

CREATIVE SERVICES/PROMOTION

The creative services department, sometimes called promotion or marketing, is responsible for promoting the station—its image, personalities, news and programming. One way this is accomplished is through the production and scheduling of promotional spots (on-air promos) which promote the station's news and programming. Stations utilize a certain amount of their commercial time to "advertise" their own product on their own air to increase ratings, or viewership. Since the station's commercial rates are based in part on the number of viewers a station can deliver at any given time to their advertisers, increasing ratings is of the utmost importance. A large station might have three or four creative services producers and a person who provides the traffic department with the specific promos to be scheduled on the next day's log.

Creative services is also involved in advertising the station in other media, such as radio, newspapers, billboards or bus shelters. Smaller stations may not have a budget for outside advertising. In a large market, however, a station may even hire an ad agency to produce its advertising campaigns. The creative services or promotion director works with the ad agency's account executive and creative team to produce these campaigns.

Another important area of promotion is public relations and publicity. I'm not just saying this because this was my life for a couple of decades. Publicists encourage newspapers' TV critics, columnists and feature writers to write about the station's programs, news and on-air personalities, and work with radio producers and magazine editors to get additional free publicity for the station. They also work with the network and syndicators in coordinating press tours and satellite press interviews, and often serve as station spokespersons when the TV station itself is in the news.

Public relations encompasses publicity and more. It may include staging premiere parties for new network or local shows, writing and producing station newsletters, giving tours and hosting special groups (we had lots of TV executives from foreign countries visit our station), coordinating sponsorship of community events (e.g., food drives and festivals), and managing a speakers bureau for on-air news folks.

The titles vary from station to station: public relations director, media relations director or publicist. A large station might have two or three handling PR and publicity, not counting interns. At a small station, all PR and publicity might be the responsibility of the creative services or promotion director, who also produces the promos.

COMMUNITY RELATIONS/PUBLIC AFFAIRS

This department, which may be comprised of only one or two people and an intern or two, is the station's link to the community, particularly the nonprofit world. The public affairs (sometimes called community affairs or community relations) director or coordinator schedules PSAs, some of which he or she may actually produce, promoting nonprofit community services, fundraising events for nonprofits and messages of benefit to the viewers. PSAs include all the "anti" messages—drugs, smoking, child abuse, unprotected sex; and the "pro" messages—prenatal care, United Way, volunteer, stay in school; as well as the spots inviting you to participate in walkathons, benefits and so on. In the past, public affairs departments regularly produced public service programming which addressed the concerns of the community and special groups. Some still do. Rosy Chu, Director of Community Relations at KTVU, the Fox station serving the San Francisco Bay Area market, produces and hosts "Bay Area People," a weekly half-hour discussion program. She also produces public service campaigns and public affairs specials.

Like most community relations directors, Rosy is responsible for filing reports with the Federal Communications Commission to insure that the station retains its FCC license, up for renewal every five years. Rosy files the problems and issues report each quarter, which tells the public what the station is doing to serve their interests, and she maintains a public inspection file at the station for interested citizens to inspect.

In smaller markets, the public affairs functions might be combined with other responsibilities in promotion, programming or news.

DESIGN/ART

The design, art or graphics department, headed by a design director, develops and implements the overall visual image of the station, using the latest in computer graphics technology.

The department's primary responsibility is providing on-air graphics for the daily newscasts and any of the station's local programs, promotional spots or PSAs. In a large station, there may be as many as eight to ten graphic artists on staff, primarily serving the needs of news. News shows utilize graphics to help illustrate individual stories. If the story is about gun control, you'll usually see an image in a box (box graphic) of a gun over the anchor's shoulder, or if it's about an individual, you'll see a box graphic of the person in the news. These graphics, as well as full screen maps and charts, are created in short order on sophisticated computer graphics equipment. They're stored electronically and a control room technician calls them up for use during the broadcast. The designers also create show opens, bumps (the transitional graphics going in and out of commercial breaks) and billboards (full-screen graphics with sponsor logos).

In many stations, the artists also design and build sets, props and signage. They often also design print ads for the station, as well as collateral pieces for sales, programming, news, promotion or public affairs. Examples would be the design of posters and flyers for station-sponsored events, sales brochures to support the sales efforts, newsletter design and giveaway items such as coffee mugs or pens for the promotion department, and invitations to station events, perhaps a televised community awards dinner or a premiere party to launch a new show.

ACCOUNTING

The accounting department is responsible for management of the company's finances. This includes not only accounting for current operations, but also the coordination of budgetary planning, forecasting and cost control. Typical jobs include managers for credit, accounts receivable, accounts payable and payroll. Because of the specialization in this department, there's little mobility from accounting positions to other areas in the station.

DATA PROCESSING/MANAGEMENT INFORMATION SYSTEMS

The data processing or MIS department assists all the departments with their computer systems and needs. In a large station with a variety of computer systems, from the news computers to graphics computers and personal computers of all kinds, the department may have three or four employees. In a smaller station, these functions might be handled by someone in engineering or news.

HUMAN RESOURCES/PERSONNEL

Many stations have a one-to-three-person human resources department which posts job openings, prescreens applicants and coordinates internship programs. It's usually a human resources coordinator who sends out those unpopular responses to resumes, often a form postcard reading, "Thank you for your interest, but we have no openings at this time that match your experience."

Human resources personnel manage employee benefits, (medical coverage, pensions, 401[k] plans), and staff development and employee relations, which could include coordination of station parties and quarterly meetings. They make sure the station complies with Equal Employment Opportunity Commission (EEOC) and FCC employment regulations and any other applicable laws. The details of layoffs and firings also fall in their court, and, in this day and age, they often work with legal firms hired by the station to deal with lawsuits. Lawsuits arising from news coverage or from copyright challenges in the programming area would involve managers in those departments; however, lawsuits by employees or former employees for sexual harassment, discrimination or wrongful termination usually involve the human resources staff working with the station's legal firm.

If a station doesn't have a human resources department, these functions are handled by someone in the administrative area.

OFFICE SERVICES/MAILROOM

Shipping and receiving functions are handled by the mailroom. Today, many mailrooms are called office services, and the office services manager often handles ordering stationery, supplies and some equipment, perhaps the company's copy machines. This is a very small, but essential, department.

ADMINISTRATION

Herein dwells the big cheese—the general manager—perhaps a station manager as well, and their executive assistants. The buck stops here. The overall success of the station ultimately rests with the general manager, who is in close contact with the station's owners, corporate headquarters, the network and, of course, the station's department heads.

TEN ENTRY LEVEL JOBS

- *Receptionist*
- *Administrative Assistant/Secretary*
- *Sales Assistant*
- *News or Production Assistant*
- *Traffic Coordinator*
- *Mail Room/Office Services Clerk*
- *Audience Services Representative*
- *Audience Coordinator*
- *Page*
- *CNN Video Journalist*

Entry Level Jobs

Job seekers in popular industries like television are well aware of the old catch-22—"I need experience to get into TV, but if no one will hire me, how can I get any experience?"

The best way to get experience is through an internship. Almost without exception, however, you need to be enrolled in college and receiving credit for your efforts. Also, internships rarely pay. Because of their importance, internships have a chapter all their own (chapter 6). Ditto, volunteering (chapter 18), another avenue to experience.

But let's say you're out of school, you have no TV experience but you have bills to pay and you want to make a living in the business. Here are your best bets for entry level positions requiring no previous television experience. I'll also critique their potential as stepping-stones to other television careers.

SECRETARIAL/ADMINISTRATIVE ASSISTANT/SALES ASSISTANT

Most department heads have a secretary or administrative assistant. You generally find one, occasionally two, in news, programming, creative services, operations, accounting, human resources and administration, depending upon the size of the department. These positions, not surprisingly, require secretarial skills, some mastery of computers and word processing, and presumably English: composition, grammar and spelling. The sales department usually has a sales assistant position for every two or three AEs. In addition to the skills already listed, these positions require working with numbers and contracts.

Most of these positions pigeonhole you as an administrative assistant. Breaking out requires a combination of initiative and luck.

The initiative will be the effort you put into learning other jobs in the station and taking on additional responsibilities that help you gain experience. Occasionally, you can incorporate this learning into your actual job by asking for more responsibility. Otherwise, you will have to stay after hours to "apprentice" by hanging out with or volunteering to help other overworked employees.

The good news about the TV business (and probably a lot of other industries today) is that the employees are often swamped with work and welcome help from someone who's bright and enthusiastic, especially if s/he doesn't require a lot of supervision.

The luck aspect of being able to move on has to do with being in the right place at the right time. Many a sharp secretary has been asked to fill in when someone else in the department has given notice and left. The supervisor is much too busy to fill the position within the couple of weeks before the quitter leaves, so as a stopgap measure, the supervisor has the quitter train a secretary to fill in until s/he has a chance to interview and find the best candidate for the job. Often a temp is brought in to help with the secretarial chores, freeing up the secretary to learn the new job. (Read more about temping in chapter 19.)

The boss, overwhelmed by the pressure of other responsibilities, often doesn't get around to posting the job and interviewing for a few months. In many instances, the secretary ends up getting the job. Even though she (in rare cases, he) has only a few months' experience, she's had a chance to prove herself and, at that point, may well be the best qualified person for the job. She already knows everyone in the station, has actually been doing that exact job for a while, and she and the boss have had a chance to check each other and the job out.

Some departments have more promise for promotion than others. In my 18 years in promotion/creative services departments, 95% of the people who started as the creative services' administrative assistant moved on to bigger and better TV jobs, often within a year. Of the dozen or so people I worked with in that position, many of whom I was responsible for hiring, three or four became promotion producers, a couple became publicists, one's a newswriter/producer, one's a news anchor on San Francisco's BayTV newscast, one's in charge of publicity for KCBS Radio in San Francisco, one moved on to the design department as an artist, and one ended up the creative services director at San Francisco's ABC O&O. I myself started out with some secretarial responsibilities in a combo secretary/publicist position at KGO after four years as a secretary in the ad agency that produced KGO's ad campaigns. After about six months, I became a full-time publicist/listing editor (providing the program listings to *TV Guide* and the newspapers), then manager of press information at KGO, then moved to KRON as director of public relations, and, ten short years later, was promoted to director of marketing services.

Your chances of moving out of this kind of entry level job are discussed in greater detail in chapter 8.

RECEPTIONIST/SWITCHBOARD

The receptionist or switchboard position is a good way to familiarize yourself with the overall operation of the station. Your ability to advance will depend in part on the market size. I once met a woman at a Broadcast Promotion Association convention (now PROMAX) who had been the receptionist for a network affiliate in San Antonio, the 42nd market. When the promotion director left without giving notice, she was promoted into the job with no experience whatsoever. And just a few months later, she was attending the national convention for TV and radio promotion people, learning all she could about the job she'd been handed.

Of the receptionists I've known, one ended up on the news desk as an assignment editor; one is now the station's viewer liaison handling viewer comments, requests and complaints; and a third became the administrative assistant in the operations department.

NEWS ASSISTANTS/PRODUCTION ASSISTANTS

The paying entry level jobs in news and programming are production assistants. News sometimes calls them news assistants. These positions usually require no previous experience in broadcasting, so many an aspiring college graduate with a degree in broadcasting or journalism applies for the occasional opening.

The challenge is competing against many other equally enthusiastic go-getters, some of whom may have already proven themselves as interns within the station, and many others who may have newsroom experience in other markets.

Darryl Compton, the KRON associate news director who hired all the interns and production assistants for more than a decade, was impressed with the number of applicants for the news assistant positions who had previous TV experience. They were eager to move into the fifth largest TV market, even if it meant taking a less prestigious position. However, because salaries in larger markets are proportionately higher than those in small markets for the same job, applicants might have to take a pay cut.

Conversely, little fish in big station newsrooms often move to smaller market stations to become bigger fish in those ponds. Although they might not be making any more money or they might even take a pay cut, they can gain experience and knowledge in more responsible, higher-ranking positions, as producers, writers, or even reporters, anchors or news directors.

So production assistant positions are hard to come by, but landing such a position will give you a chance to learn a lot about the business, discover if it's for you, make valuable contacts, and be in a position to move up. Most TV journalists have started out in this position.

TRAFFIC

This is one of two departments in TV which have traditionally provided a gateway to a broadcasting career. Although traffic jobs are clerical and pay is low, you have a lot of responsibility putting together the daily traffic log. Your job requires tremendous accuracy and attention to detail, and if you're good, rather than promote you, your supervisor may well want to keep you doing what you do so well.

You also don't have much contact with news or programming and you won't learn much about the more creative aspects of the business, unless you tack a few extra hours onto your day and develop some kind of apprentice relationship with a mentor in another department.

The most logical opportunities for advancement would be within the traffic department to a supervisory position, or into other departments such as operations into positions that require working with numbers, schedules and computers.

OFFICE SERVICES/MAILROOM

Starting in the mailroom is the other traditional entry into television, largely because the positions usually require no experience. Old timers in the biz love to brag about starting in the mailroom. Their stories always have that wonderful Horatio Alger, rags to riches theme. One of my former general managers started in the mailroom.

My favorite mailroom tale is Jim Gaughran's story. After graduating with a degree in English literature from Stanford and journalistic aspirations, he finally landed a job in the circulation department of the *Times Tribune,* a now-defunct Bay Area daily paper. His job amounted to dropping off bundles of papers on street corners in the wee hours of the morning for the paper carriers. Disappointed with his new career in journalism, after nine months he headed to Idaho to help his dad build a house. A year later, with $2,000 in his pocket and a resume, he was back in the Bay Area, where a friend at the paper recommended he look for a career in television. She thought the pace of TV better suited his personality. Here's how he relates his job search:

"I called three stations, got their addresses and drove up to San Francisco from Menlo Park. I went to KPIX first because it was near the off-ramp. They didn't even have a job listing at the front desk, but I dropped off my resume anyway. At KGO, I found job listings in the lobby, but nothing that I even recognized as employment. They all sounded like very technical positions. I had no idea what I was doing, so decided to go home. On my way to the freeway, I ended up driving by 4 [KRON]. The only job listing I recognized was for the mailroom and I thought, 'I could do that.' I called the mailroom supervisor from the lobby. He gave me an interview on my 25th birthday."

Jim and the mailroom super hit it off, comparing recent vacation experiences in Mexico. And within just a couple of days, Jim had broken into TV. It was 1981. By 1991, Jim was producing the CBS Morning News in New York. I'll share how he got there in chapter 27, Once You're In.

ENGINEERING

Breaking into TV in the engineering/operations area is particularly difficult. Large stations are union shops, and the engineering staff is made up of permanent employees, many with years of experience, and a regular pool of experienced on-call temps. The shrinkage of local programming production at local broadcast stations, combined with technological advances such as robotic cameras, have reduced staffing needs in engineering.

Glen Kinion, who is now a video editor and shooter at San Francisco's NBC affiliate, got his break about a decade ago while studying at Santa Rosa Junior College. He landed a technical internship at a new independent station, Channel 50, only 20 miles from his home. He says, "I ended up in the news department and learned everything on the job. I learned to shoot there. Because it was a small market, you had to do it all. I was allowed to create my own shift, and ended up putting in 35 hours a week, four or five nights a week, exceeding the internship requirement." Glen was allowed to intern for two consecutive semesters, and as he sees it, "I traded two semesters of free labor for a career." The station started hiring him for vacation relief, for which he earned "a whopping $50 a day." But today, he's in the fifth largest market, with steady, albeit temp, work doing what he loves.

Most of the techs who get into TV today have electronic experience in some other field. Breaking in is usually easier in a non-union station, which means a smaller market.

MEMBERSHIP OUTREACH

This is a typical entry level area in public broadcasting, at your local PBS station. Attracting and keeping members is essential in "member-supported" television. The membership outreach department coordinates mailings and telemarketing to encourage members to renew.

AUDIENCE SERVICES

Many PBS stations have audience services representatives who take calls from viewers, record their comments and answer their questions. (Some of the larger commercial stations may have a person assigned to this job, although most commercial stations filter the questions to individual departments, such as PR, news or programming.) At WGBH, Boston's PBS station (which has an astounding 1,000 employees in television and radio, including their temp and freelance staff), there are 10 audience service reps. Because WGBH produces 33-35% of PBS's prime time schedule, it falls somewhere between a local PBS affiliate and a network. Shows produced at WGBH for the entire PBS audience include "Nova," "Frontline," "American Experience," "This Old House" and "Victory Garden." "Masterpiece Theatre" and "Mystery" are two PBS series WGBH coproduces with British television.

Michael Scannell, one of the ten audience service reps at WGBH for the past two years, says the single most asked question he gets is "How do I get my house on 'This Old House'?" He also gets all kinds of obscure questions, some about a clip of music on a show. I spoke with him the day before Thanksgiving and he'd received three calls for recipes featured on Julia Child's shows ten to fifteen years ago! During on-air pledge periods, the entire development department, which includes the ten audience service reps, gets involved. Scannell serves as team leader, training volunteers and deciding which premiums (give-aways to encourage new members to join at higher levels) are working. Because he has a radio background, he also gets involved in WGBH's radio pledge programs.

CABLE INSTALLERS AND CUSTOMER SERVICE REPS

At local cable systems, the entry level job on the technical side is installer, and in administration, it's the customer service representative. The operation of local cable systems and a variety of career opportunities in all areas are discussed in detail in chapter 23, Careers in Local Cable.

CNN'S VIDEO JOURNALIST PROGRAM

CNN's entry level program at their Atlanta headquarters is described in great detail in chapter 24. If a news career is your goal, CNN's VJ program is excellent.

The TV Aptitude Test

You've read about the business. You've decided it's right for you. So now we arrive at the moment of truth: Are you right for it? Do you have what it takes to work in TV? See how many of these ten basic qualities you possess.

DESIRE

Do you feel you simply *have* to work in this industry? Those that do usually make it. Although desire alone will not get your foot in the door, there are very few people who get in without it. Desire is power. It's the fuel that drives your job search. It provides the persistence and stamina necessary to overcome the competition.

ENTHUSIASM

Real estate ads boast "location, location, location!" Selling yourself to a TV manager requires "attitude, attitude, attitude!" In every interview I have conducted, a great attitude was the first attribute most TV professionals were seeking in entry level employees. This includes not being too good to answer the phone, open the mail or make photocopies. The ones who make it are the ones who convince the hirers that they will do anything: sweep the floors, make the coffee, pick up lunch. It's a rare TV bird who hasn't paid his or her dues. And because everyone's done it, it's expected of the next generation.

INTELLIGENCE AND RESOURCEFULNESS

One of the greatest advantages of working in television is the stimulation of working with bright, creative and energetic people. To successfully compete for entry level jobs, you must convince the hirer that you're the best and the brightest. First, you need to be resourceful even to get the interview. Once you're hired, you need to learn the job quickly. Many supervisors, particularly in the news and engineering areas, are faced with continual pressure—getting tape edited, stories written—and have little time to train the new kid on the block.

WRITING SKILLS

Although writing skills aren't required for technical positions, they are absolutely essential for a career in news, programming, creative services, public affairs, sales and management. The greater your command of the English language—spelling, grammar, vocabulary—the bigger an asset you'll be to the station. Are you a born communicator? If you like to tell stories, you'll feel at home in TV news or programming.

COMPUTER AND TECHNOLOGY LITERACY

Computers are being used in all areas of television today. The future belongs to those who know what to do with them. In news, the advantage goes to those who know LEXUS, NEXUS, Newsstar, and BASYS—how to access information through the myriad of online information resources. The applicant who's mastered not only word processing but also spreadsheets and other software is more likely to land the administrative assistant job. And with the state-of-the-art technology in engineering, the winners are those who can quickly adapt to new equipment, or even occasionally repair a problem without calling in the maintenance technicians.

FLEXIBILITY

No more 9 to 5. If you're the kind of person who's obsessed with working out at 5:15 every day, or resents having to put in an extra five or six hours on short or no notice, TV is not for you. TV is not a career, it's an obsession. While it's true that employees in some departments (administrative, sales, creative services, human resources) tend to work more regular shifts, working more than 40 hours a week is common in the business. Of course, the nature of news makes it a round-the-clock proposition, and if you're low person on the totem pole, you could well start out working weekends or the graveyard shift. Because stations want to get maximum use out of their facilities and equipment, producers in programming, public affairs and promotion and their crews are often scheduled for production or editing in the evening.

ENDURANCE

When major stories break, you may be called upon to work all night even if you aren't in news. Although breaking stories primarily involve the news staff, all station employees are expected to pitch in wherever needed. After the Loma Prieta 'quake of '89, KRON provided live news coverage around the clock for a couple of days. Some non-news personnel helped out in news, others went out to find flashlights and food in a city with no power. I searched for phones in the building (that worked and weren't already in use) for Bryant Gumbel, Jane Pauley and their producer. Then I headed out to rent a truck for "Today" to haul their set and equipment around. The day the U.S. starting bombing Iraqi forces during the Gulf War, I was assigned to edit news wire stories from 2:00-5:00 a.m. to feed to viewers with closed-caption boxes. Inexperienced on the BASYS news computer before, I had a half-hour of training that day. It wasn't the first time I brought in a sleeping bag and cot to catch a couple hours of sleep between shifts—helping out in the middle of the night doesn't excuse you from your day job. Even when breaking news isn't the culprit, programming producers and editors occasionally work through the night to get a show on the air.

MOBILITY

Your willingness to go where the opportunities lie greatly enhances your chance for success. You'll read about the advantages of getting your start in a small market in chapter 7. The normal progression of media careers is moving from market to market—either to a smaller market for a bigger job with more responsibility, or to a larger market station to the same job or even a lesser job. A news director from a small market may step back to become an assistant news director or producer in a larger market, and make more money in the bargain. But do keep in mind, it's not about money. It's about learning the trade, making contacts and broadening your horizons.

MULTIPLE TALENTS

The wave of the future is having one person do three or four jobs simultaneously. It's always been the case in the smaller market newsrooms, but now even the top ten markets are hiring individuals who can write, edit, shoot, field-produce and report their own stories. And even if you don't have to perform all the jobs yourself, the more you know about the skills required of the rest of your team, the better you'll be able to do your job. Even outside the news, programming and engineering areas, employees are being called up to fill multiple roles. Jobs are being combined, and those who can wear several hats at one time are getting and keeping the jobs.

CURIOSITY

This is essential for anyone who's involved in any aspect of gathering news, uncovering facts, determining the who, what, where, when and why of events and issues. The greater your curiosity, the more you'll dig to get to the story behind the story or to the deeper significance of an issue. News directors don't want assignment editors, producers, writers and reporters who accept at face value what they're told by a company spokesperson, a public servant or a lawyer. They want all sides of the story. Curiosity doesn't just apply to news. Curious people ask questions, look for better ways to do their jobs and want to learn about the latest technology.

OK. You say you've got what it takes? It's time to get started.

TEN MORE QUALITIES FOR A SUCCESSFUL TV CAREER

- *Reliability*
- *Creativity*
- *Good communicator/listener*
- *Team player*
- *Ability to focus*
- *Quick study*
- *Sense of humor*
- *Attention to detail*
- *Confidence*
- *Resilience*

Preparing for the Break-in

Getting Started

Launching a career in TV may seem overwhelming, especially with all the talk of intense competition, but you have to start somewhere. Every journey starts with a single step.

When I got the notion that I wanted to break out of my secretarial rut in the ad business and get into TV, I took a one-day seminar at U.C. Berkeley entitled, "Opportunities for Women in Television and Radio." I was discouraged by the emphasis on competition and let an entire year pass before I got my resume together and began my job search. Once I finally started interviewing, it was only a couple of weeks before I'd gotten my break and was working at KGO. The moral of the story is, your chances of breaking into the business are much greater if you actually *try* to get in.

If you're still in college, Professor John Hewitt of San Francisco State University's Broadcast and Electronic Communications Arts department recommends that you start working on your job search three to four months before you graduate. He says you must set aside some resources for postage and phone bills and start that essential process, networking.

Think of your job search as a full-time job. It's not a hobby, it's your life! Set goals. Create a timeline. Target your market. Develop your strategy (read chapters 7 and 8). Be systematic and organized. Your plan will include research, developing your resume and cover letters to be adapted to various situations. Your tools will be a word processor or computer and decent printer, a phone with a message machine, a calendar to plot your attack and serve as a tickler for follow-up phone calls and a filing system to keep track of conversations.

Bill Groody, president of North Country Communications, which owns five radio stations in Northern California, said to those wanting to break into television and radio at a 1992 Radio-Television News Directors Association regional convention, "You have to have a strategy. Start by figuring out what you want to do. Do you want to be a reporter? Do you want to be a producer? Do you want to be a news director? The earlier you can set that goal, the better off you're going to be."

So the question remains. Where do you start? And the obvious answer is, you start from where you are. Elena Nachmanoff, vice president, talent development, for NBC News, recommends, "Make use of the school that you go to and your knowledge of your hometown. Go home for the summer. Even if you have to work during the week, perhaps you can volunteer on Saturday and Sunday at a local station." Jim Gaughran, news producer at KRON, agrees, "Apply where you live if you like it there. Use the fact that you know the area as part of your pitch. If that doesn't work, you have to go someplace smaller."

Actually, you're on the right track if you're reading this book. When you get to the end, you'll have a leg up on the competition.

Education

Although you may find some entry level jobs which don't require a college degree, you won't get far in today's competitive market without one. If you've already got your degree and it's too late to take advantage of the advice in this chapter, don't panic. Experience, attitude and contacts are also important. If you've got college ahead of you, I'd put money on your success. You're obviously taking the initiative and planning now for your future. That's a great start.

HIGH SCHOOL

Very few TV stations have internships or any kind of summer or temporary work for high school students. A few do, however, and you won't know if you don't call. So find out about any opportunities to volunteer or intern or, better yet, work for money during your school breaks or on weekends. Also, take advantage of any job-shadowing programs or career days your school offers.

Don't hesitate to call people in different departments at the station. You can cold-call, but it helps if you know someone who knows someone so you can say "So-and-so gave me your name."

I was always impressed when I got calls from bright, mature high school students who were curious and enthusiastic about the business. I always invited them to our career panels, and if they asked nicely, I'd even allow them to come down and spend part of a day at the station, watch a live newscast or sit in on the morning news meeting where producers and managers decide what should be included in the day's newscasts. If I was really impressed with them, I might even give them a tour and introduce them to producers, editors, shooters and technicians along the way.

Tours at local stations are becoming more difficult to come by, since stations rarely include this responsibility in anyone's job description and staff are overloaded with other duties. However, certainly anyone you reach at a station can show you around if s/he is so inclined. Rather than asking for a tour, which could take a half-hour to an hour of someone's time, start by just asking if you can watch the news from the control room. This is more interesting than watching from the studio because you see all the behind-the-scenes action. It also requires less time on the part of your host, as you can be left in the control room for the duration of the show. Once you're in, maybe you can talk your host or someone else you meet into a quick tour.

While at the station, be respectful of people on deadlines, but if you are introduced to employees, or a couple of them strike up a conversation with you, have a couple of good questions ready for them. What do you do? Do you like your job? Why? What kind of education or background do you recommend?

Oh, yes, and while in high school, do your homework! This is a great time to develop good research habits, learn to write and perfect your English. And pay attention to history, government, science and social studies. They'll all be invaluable later on—in college and after, no matter what career you choose. Also, take advantage of any and all computer training you can get.

Michael Scannell launched his broadcasting career while still in high school, demonstrating how to find a need and fill it. He put together a demo tape, using his high school facilities, and approached a little Boston AM radio station. He said, "I'll make a deal with you. Saturday afternoon when all the high school football games are over, I'll call up the local coaches and get their comments about the games. I'll put it together in a 3-minute feature and feed the package to you over the phone at 5 p.m. each Saturday." For two years, the enterprising high school student made $50 a week putting together that radio package.

Scannell went on to study mass communications at the University of Hartford, landing radio internships in Boston. After college, he worked for a couple of radio stations before starting a recording studio. For five years, he produced radio jingles and commercials. After a stint in the insurance business as a customer service rep, he went after and landed his first TV job—as audience service rep for WGBH.

CHOOSING A UNIVERSITY OR COLLEGE

There are more than 300 American colleges and universities offering degrees in broadcasting and communications. A listing of colleges with broadcasting programs is available through the Broadcast Education Association, 1771 N Street, NW, Washington, DC 20036. Or take advantage of reference materials at your high school or public library.

The school you want depends on what *you* want. Some schools are known for their journalism departments, others for their broadcasting departments, some for neither. Call up TV stations in your area, ask for the directors of departments that interest you—news, engineering, programming—and ask them which schools they recommend in the area, or even across the country. Having questions about education is a great reason to call people in the business and maybe make some contacts which will help you when you're looking for internships or job opportunities later on. While you're on the phone, you can find out more about their operation and availability of internships, part time work or volunteer opportunities.

Professor Hewitt of San Francisco State says, "You want a school with a 50-50 blend of theory and practice, and a good internship program, preferably in a large market. A school where the teachers have worked in the field, or written the principal textbooks that are used. The balance of theory and practice allows you to branch out. If you're not happy with your first choice of areas, you can choose something else. You end up with transferable skills."

If you choose a school where you get good hands-on production experience, you'll get a chance to produce videos, news stories and programs that prospective employers would like to see.

Some of the best training is acquired at schools that, like the University of Missouri, have their own radio and/or TV stations. As a student at San Jose State College (now a university), Valerie Coleman was "discovered" in 1968 when her coverage of student protests on the college TV station ended up airing on local Bay Area stations since the college cameras were the only ones to capture the story on film. Valerie went on to report and anchor for television and radio in San Francisco, Los Angeles and, currently, New York City.

Even if you aren't discovered, you'll be able to put together a resume tape based on work at the college station. While at the University of California at Berkeley, Shari Jackman, whose goal was to be a sportscaster, was a sports anchor and play-by-play announcer on the college-run KALX Radio for three years. She found it very helpful because "it gets your confidence level up, particularly if you want to be a reporter. You get the practical experience of going out and conducting interviews, of being on the air."

Choosing a college or university in a large television market means a greater variety of potential internships. For example, San Francisco State offers internships at 60 different local TV stations, production houses, cable stations, even at George Lucas' Skywalker Ranch and LucasArts.

If you intern in a large market station, there will also be more graduates from your alma mater working in the local stations, thus increasing your network of contacts in the business.

CHOOSING A MAJOR

The consensus among the television hirers I've interviewed is that a broad liberal arts background is ideal for most careers in television.

Professor Hewitt says, "For news, it's very important to finish your undergraduate degree—in humanities, social sciences or sciences. You need the ability to organize, research and interact with people. Get any degree that you're interested in, government, international relations, history. It's not imperative to get a broadcasting degree. If you do decide to major in broadcasting, be sure to take a variety of other subjects."

Before settling on a major, find out about the school's internship programs. Some schools' TV internships are only open to those in communications, journalism or broadcasting. You may be able to find a way around the "rules," but you're better off knowing the rules going in.

Mike Gaynes, a newswriter, reporter and anchor for the past 15 years, says his journalism degree from the University of Oregon gave him "a certain knowledge of basic principles involved." He feels that you *can* learn to write in college. And he adds, "You have to have that walking in. If you can write, you can work. But you need history, political science, law, to give you broader knowledge."

Professor Hewitt advises, "While in school, don't pass up the opportunity to use their equipment to produce news stories, programs, etc. You'll be able to use all of this in your resume tape, and the equipment costs a lot more to use after you're out of school. [See chapter 22, Resume Tapes.] Also, take advantage of the criticism you get in school. Don't bristle. It's the only place you'll get criticism and *not* get fired. In a small market station, they don't have time to train you, they'll just let you go, so listen to the criticism you get in school."

Joe Fragola, executive producer of BayTV, a 24-hour cable news and information station in San Francisco, recommends taking business and marketing classes. He says, "I'd take a finance class to understand the business. I'd take communications classes, but communications would be my minor." In fact, Joe graduated in 1970 from a small men's college in Rochester, New York, with a degree in political science and English. Because of his love of old movies, he enrolled in graduate school at Florida State, where he says he "got the bug" and went after a creative Master's degree. He says, "I learned how to produce, I learned how to write, but I never took a traditional journalism course."

Many colleges have associations for students interested in broadcasting or student chapters of national organizations that foster excellence in broadcasting and journalism, such as the Society of Professional Journalists (SPJ), Alpha Epsilon Rho and American Women in Radio and Television (AWRT). These are great places to network with like-minded students and learn more about the business.

TECHNICAL EDUCATION

Jim Owens, director of engineering at WLS in Chicago, feels a year or two of a technical school, such as DeVry, which has schools across the country, is sufficient for a career in engineering. He feels a four-year degree in communications might be "a little excessive" to be a master control operator. A lot of junior colleges also offer two-year degrees in electronic technology.

A number of successful engineers today got into the business while they were still in college and left school before earning their degrees. Daniel Murphy, now an engineer at KRON, was a business major at San Francisco State when a job as a still photographer for *The San Mateo Times* fell into his lap before his 20th birthday. After one more year of juggling school and work, he opted for work. Daniel ended up in the fast lane, moving over to the *Oakland Tribune* as a photographer, traveling a lot, covering Super Bowls and the President. He was 26 when he decided to break into TV, landing a temp job at KRON. About getting a college degree, he says, "Now so many people want to get in, it doesn't hurt to have that degree, just in case they ask for it. If an employer has a choice, that degree shows, if nothing else, that the person can start something and finish it." Daniel plans to go back to college for a fine arts or photography degree.

Glen Kinion also left Santa Rosa Junior College to work full-time. He regrets not having earned his degree but says it hasn't hurt him yet. "I ended up with a pretty good background in terms of basic journalism and ethics by studying the people I worked with. I had to pick up knowledge of the workings of state, local and national government on the job."

POSTGRADUATE EDUCATION

If you're out of college, you can enroll in adult education classes or your community college to gain more skills that may help in your pursuit of a TV career.

The Radio-Television News Directors Association (RTNDA) offers continuing education, and in this age of the information superhighway, more and more education is available online.

In larger markets, local nonprofit organizations offer classes for those who want to learn more about the business. One such Bay Area organization is Media Alliance, with classes in newswriting, copy editing, computer skills, marketing oneself and much more. The basic $60 annual membership includes their monthly publication, *MediaFile,* which talks about what's going on in the local media scene. And their job file lists hundreds of openings, not only in TV but also in ad agencies, production companies, corporate and cable. Members can access a job line and pay extra to have listings mailed to them on a regular basis. Find out if there's such an organization in your area.

On two occasions, furthering my education furthered my marketability. When I was trying to break into the advertising business at age 23, the entry level secretarial position I was seeking required shorthand. I convinced the boss that I was a very fast writer *and* I would learn shorthand immediately. I enrolled in night school, on my own time and at my own expense, to get the skills I needed to land that job.

Before my interview at KGO for the publicist position, I heard that the boss, Dennis Fitch, was looking for someone who could write feature articles in the entertainment section of the Sunday *San Francisco Chronicle.* I had no experience writing features, but found out about a U.C. Extension course in feature writing taught by a *Chronicle* writer, no less. I told Dennis that although I had no doubt I could handle the writing aspects of the job, I was going to fine-tune my feature writing by taking the U.C. course at night, which was starting in just a week or two.

These kinds of gestures not only improve your qualifications for a job (or help you meet the minimum qualifications), but also let you show the prospective employer that you are willing to do *whatever it takes* to get the job and then get the job done.

Graduate degrees will definitely give you an edge in the management and financial areas of the industry. In other areas, some hirers are impressed by graduate degrees and feel the added knowledge may make you a more valuable employee. But others might feel your graduate degree makes you overqualified for a bottom-of-the-rung entry level job.

Internships

WHAT IT IS

An internship is an opportunity for a student to work for a TV or radio station, production house, film studio, cable operation or network, while receiving college credit for the experience. Most internship programs require a commitment of 15-20 hours per week, although supervisors are usually flexible. Internships usually run three or four months, coinciding with the fall semester, the spring semester and summer. Some, such as those at CNN, are quarterly. (See chapter 24 for more on CNN internships.) Some stations pay a $5-per-day stipend. A few even pay minimum wage. The majority of them are nonpaid.

The departments that use interns vary from station to station, from supervisor to supervisor, and from semester to semester. Some stations accept interns only in their news departments, others may use them all over the station. KGO-TV, San Francisco's ABC affiliate, offers internships in news, sports, programming, public affairs, finance, graphics, promotion and research. Other departments in which individual stations might offer internships are engineering, sales, weather, data processing and personnel/ human resources.

BENEFITS TO THE STATION

The bottom-line benefit to the station is free, or at least cheap, labor. Yes, interns do require training, but many stations rely on sharp interns to perform tasks they normally would have to pay for. In news, interns sometimes do the work of desk and news assistants; in promotion, publicists; in programming, production assistants. Many departments delegate administrative assistant responsibilities to their interns—answering phones, photocopying, filing or handling correspondence.

In addition to saving on salaries and employee benefits, the station has the opportunity to discover new talent. Some interns show tremendous promise and prove to be excellent candidates for entry level jobs, although the station may have to wait until you graduate.

By the time your internship is complete, you may already be trained for the job. You already understand how the station operates, can find your way to the mailroom and know the players and the station's programming. After an internship, the supervisor has had a chance to measure your intelligence, initiative and attitude. How did that attitude hold up through boring assignments? How much supervision did you require? How quickly did you catch on?

BENEFITS TO YOU, THE INTERN

You can learn as much as you want. But it may require a measure of personal initiative. In some cases (in a nonunion shop), you can get some hands-on experience on the latest equipment. You can learn to use the news computer and other information systems. You can find out how the TV station operates, make a lot of excellent contacts for the future in a variety of departments, and add some impressive experience to your resume. Most importantly, you can determine if you want to pursue a career in TV, and how badly you want it.

Jim Gaughran says, "An internship gives you an opportunity to look at the business. It looks glamorous on the outside but it's all pro wrestling on the inside. It's messy, there are huge egos, difficult people, strange people, extreme deadline pressure. It's really not for everybody."

Diana Mordock, who went back to college in her mid-thirties to get a broadcasting degree, landed a news production assistant job at KRON— but soon realized it wasn't for her. In retrospect, she says she feels internships should be mandatory for broadcast majors, so they'll know what to expect before they expend a lot of effort trying to get in.

REQUIREMENTS

These will vary somewhat from station to station, not only in the stated requirements for internships, but also in the degree of diligence with which the regulations are followed.

KGO, which pays minimum wage to its interns, is very specific about requirements. You must currently be enrolled at a four-year college or in graduate school, should be a junior, senior or grad student, should be available 10-19 hours per week for the entire semester, must receive course credit and must reside in the San Francisco Bay Area during the internship. Application requirements at KGO include filling out an internship application with a signature of the school representative verifying that you're receiving credit, your current resume, a one-page statement outlining why you would like an internship at KGO, what you expect to learn, what you could offer, your career goals and interests, two letters of recommendation from college professors and/or counselors, and an official college transcript of your last semester. Whew! But then again, KGO pays! That's a real rarity in the internship game.

At most other stations around the country, a resume and a cover letter will suffice. Most require college credit, although some don't keep very good records and their personnel departments don't get around to demanding the proof. Similarly, some supervisors are a little lax about the rules and don't enforce them. I've known a number of interns who've gotten past the college credit requirement.

If you're out of college, you can check to see if a school in your area offers an internship class in which you can enroll even if you aren't a full-time student. Occasionally, a community college or adult education institution will offer that option, which might suffice at some TV stations. You may end up paying a hundred to two hundred dollars to enroll, in order to work for free, but the experience and contacts could be worth it to you.

HOW TO GET AN INTERNSHIP

In larger stations, internship applications are filtered through the human resources or personnel department. Whether or not you're called for an interview will depend on how many internships are available, how many other applications have come in and how yours compares. However, those who actually get the internships are often those who have gone directly to the department where they want to work or know someone at the station.

To be competitive, you must be proactive. Go ahead and send your resume to personnel, if so requested. But also find out who's responsible for hiring interns in the department in which you'd like to work. Talk to that person on the phone, ask if you can drop by with your cover letter and resume, and say hello while you're there. (Maybe you can turn it into a mini-interview.) If not, at least send a copy of the cover letter and resume directly to the person making the decision. If you can't get through to the decision maker, try to make an ally of that person's assistant or someone else in the department who can put in a good word for you. Read more about this in chapter 11, Networking, and chapter 12, Making Your Phone Work for You.

Don't wait until the last minute. Some smart students make contact with the station and the decision maker six months to a year before they want to do the internship, and may even get a commitment several months in advance. I've known college students who've managed to do two or three internships in different departments, lining up sequential positions months in advance.

If you're not sure where you'd like to work, do your homework. Call the station and talk to the personnel department (if they don't have one, begin with the receptionist). Find out which departments use interns. Ask to be connected to the departments that interest you and ask what the internship entails, i.e., what kinds of tasks you'd be performing. This can be tricky. Supervisors want to hire students who are enthusiastic about the internship and who convince them that no job will be too small, no task too menial. You may want to speak initially to someone other than the supervisor, so you can determine if this is an internship you want. If it is, you can call back later and be genuinely excited when you talk to the decision maker.

All internships are not created equal. And if you can befriend someone on the phone who can give you a little more inside information, you'll be ahead of the game. Some internships are more competitive than others. It's supply and demand. If you don't land an internship in your area of greatest interest, say news, try getting an internship with the manager of media relations where you may get a chance to hone your writing and reporting skills by interviewing news and programming producers about their series and specials, and then writing press releases to publicize that news product. You may even be able to talk a department head who didn't think they needed an intern into needing *you*.

One reason I recommend going directly to the department that interests you is that the human resources or personnel coordinator who receives your cover letter and resume may not pass it on to the appropriate department in a timely fashion, not out of malice, but because s/he's swamped. I've heard from several department heads who have gotten very excited about a potential intern whose resume has just arrived in their in-basket, only to call and find out that the student has already accepted another internship. Often the student gives up after not hearing for several weeks, and accepts something less desirable. So the fact that your internship application or inquiry arrived well before the stated deadline at a station, broadcast or cable network doesn't mean it will get to the appropriate decision maker right away. Don't be discouraged if you haven't heard back. Get on the phone. You might be just the person they're looking for. They just don't know it yet, because they've been too busy to get to those resumes.

If you're going for an internship at a station or production house that doesn't pay, which is the vast majority of them, you'll find the "hiring" process can be pretty lax. This flexibility will work in your favor if you're a go-getter. (And if you're not, this isn't the business for you.)

Shari Jackman, while visiting KRON with her U.C. Berkeley journalism class, took the opportunity to introduce herself to the station's sportscasters and producers and ask about internships. One of the sports producers told her to call back in five months. When she did, that producer was gone. But she persisted and sent in a resume. Despite the fact that her contact had moved on, she ended up getting a sports internship. She said, "Although it was a nonpaying job, it was heaven on earth. I watched sports all day, logging videotape, selecting game highlights and helping to edit clips. I did all of the basics." Shari was able to parlay her internship into a paying news assistant job shortly after she graduated.

MAKING THE MOST OF YOUR INTERNSHIP

In an ideal world, you, the intern, would be systematically trained in a variety of areas by kind, caring mentor figures. In the real world, however, it's just as likely you'll find yourself surrounded by people "with their hair on fire," in other words, harried, frazzled and fighting a deadline. They just don't have time to train you today, and you may get the sense that they don't want you in the way. They may even be dumping unenlightening grunt work on you—photocopying, filing and answering the phone, or worse, handling personal errands. Granted, this is a worst-case scenario, but it can happen.

In this situation, you, clever intern that you are, do all tasks quickly and cheerfully, knowing that it's all part of paying your dues. If you're going to rise to the top, you find other ways to be of service, like the Buddhist who finds bliss by sweeping the floor perfectly. You watch what others in the department are doing, and without being obtrusive or a pest, ask questions, offer to help and try to learn. You make a very positive impression when you find something that can be improved—a file updated, a common workspace cleaned up or rearranged—and you take it upon yourself to find things to do when you don't have a specific assignment.

Rosemary Wesela, who's "seen it all" in her dozen or so years as assistant to KRON's various news directors, explains the value of even the worst-case scenario internship: "You do the lowest of low work. You answer the phones, you go for coffee, you do all the things you don't want to do, but you're also there to hear the language, to understand how things are put together, to understand the detail and the time things entail. Just by being there, you learn something."

Fred Zehnder, news director for KTVU, the Fox station in Oakland, said, "Some interns come in and hone their skills to the point where they are useful employees to the newsroom. We have seven or eight full-time employees who started as interns and left an indelible impression as to how good they were. When we needed casual [temp] help, we hired them. They came in and had an aptitude for TV news. People took notice of them right away."

During your internship, introduce yourself to people around the station, ask if you can sit in, on your own time, on their tapings, edit sessions, or even production meetings. As a courtesy, let your supervisor know what you're up to. If you're doing an excellent job and taking a lot of initiative in your own department, your boss will be delighted you're learning about other areas of the station's operations.

If you've got a news internship, Professor Hewitt says, "You must learn the electronic databases: LEXUS, NEXUS, the news computer systems like BASYS. Schools often can't afford the latest in computer systems, so take advantage of what you can learn while you're at a station."

You may also have an opportunity to talk to people at stations in other markets during your internship. Take advantage of these opportunities by asking about jobs at those stations. You'll have more clout when you're an intern because you'll be working in TV. You may even have contacts or information they can use.

Joe Fragola suggests that as an intern, you "stand up above the crowd, be different. Don't bounce off the walls, but be enthusiastic, energetic, enterprising. And you've got to understand how to ask questions."

After you land that internship, Fragola says you have to ask yourself, "What am I bringing here, what am I going to learn and if they're not teaching me, how am I going to find out?"

Some interns complain bitterly that passing out memos isn't journalism, that they have already mastered photocopying, that they're bored and aren't being given exciting assignments. Big mistake. They stand out, all right, but for all the wrong reasons.

WHEN IT'S OVER

Even though everyone I've talked to over the years sings the praises of internships, landing one is just the beginning. Even if you're the best intern the station has ever seen, there just aren't very many openings for entry level jobs, and one might not be available when you're ready. Even if there is one, you may be competing against people who've had several years of experience. It could take a lot more effort to parlay your experience into a job. Use the "fan club" you've developed. Think of them as members of your team. Keep in touch with them, call them on a regular basis—every month if you're actively job hunting—and ask if they've heard of jobs through the grapevine. Perhaps they know someone at another station where you've discovered an opening. Ask them to call that friend or acquaintance and put in a good word for you.

For many a college graduate, the big market internship has been enough to land a good job in a small market.

Although it's important to be realistic, there's tremendous power in belief. And good things come to people who feel they deserve them. There's a self-fulfilling prophecy in operation here. If all the talk of competition and paying dues results in your belief that your chances of success are too slim, that very thought may get in your way. If you emanate confidence and a positive outlook, despite rejections along the way, you're in line for success.

Try to rise above the job search. Don't take it personally. Think of it as a numbers game. With each rejection, you're a little bit closer to the job with your name on it.

INTERNSHIP SUCCESS STORIES

Janette Gitler, currently KRON's director of local programming and community relations, was an intern in Minneapolis/St. Paul (11th market at the time) on a local morning talk show. She was a speech/communications major, with a minor in journalism, at the University of Minnesota.

She'd been an intern for three weeks when a job opened up for an assistant producer (a step or two higher than production assistant). She asked her mom, "Do you think they would think I was crazy if I applied for this job?" They advertised in *Broadcasting* magazine and 75 applied. She explains, "Because I was an intern willing to take on anything, I could be put to the test while they were interviewing all these other people, and I got the job. Six weeks into the internship, I became the assistant producer. Six months later, the producer left, and I practically begged them to let me produce the show while they were looking to hire a producer. A very smart move, because I ended up getting that job."

The last four people Janette has hired to work in KRON's programming area have all been interns at the station.

Manuel Gallegus started as an intern at KRON. He went to school full-time and he worked full-time. Manny laid carpet in the newsroom. Manny cleaned out the storeroom and didn't complain. Manny did whatever it took. Rosemary Wesela explained, "So Manny was hired at the station as a news assistant and was the best news assistant possible. Because whatever you needed done, he did it, and he did it happily and he did it 100%. He worked here for years as a news assistant while he was in college;

any shift. He was very highly regarded by the staff; management loved him. But Manny had to go to Santa Barbara to get a job as a reporter, even though everyone loved him, even though he worked his buns off. He did some writing here, did a little producing here. From Santa Barbara, he went to Santa Rosa. From Santa Rosa, he came back to Channel 20, a small independent station with a tiny newsroom, but at least he was in this market. Then the news director here wanted to give him a chance and gave him a reporting job at KRON—on the overnight shift, 11:30 at night until 8:00 in the morning. Manny did the 24-hour cut-ins (30-second news updates). He had the worst job in the newsroom. He didn't like working overnight, but it was a chance at a network affiliate so he took it, and he took it gladly. And he did that job for just about two years. Finally, he got taken off the overnight and worked weekends and then a couple days during the week. Just recently he was put on nights, and someone at CBS saw his work. CBS News called and now he's a correspondent with CBS, based in L.A., working on what they call 'NewsNet,' feeding news stories to their affiliates."

Choosing Your Target

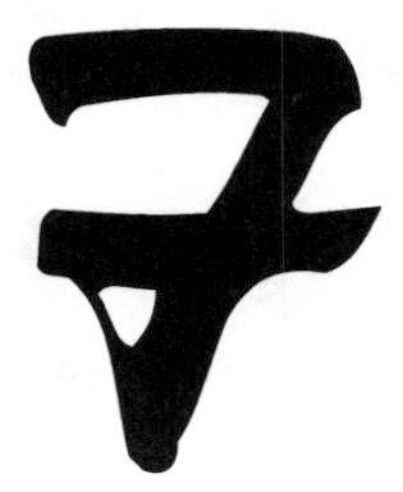

Small vs. Large Market

As you're developing your plan of attack, consider again the relative merits of small and large markets. Fred Zehnder says, "Spending a year or two at a little station is the best schooling you could ever have."

Your smaller markets are great training ground whether your aspirations are on-air or behind-the-scenes. If you're interested in reporting or anchoring, you not only will have to go to a small market to get a job, but you'll want to. You make your mistakes in a small market. In a big market, too many people will see them. Besides, in the bigger markets, the on-air jobs are going to go to those who have experience, who can start the first day and go out and cover a story. In a small market, they expect to train you. In a large market, they expect you to hit the ground running.

In a smaller market, you're going to learn how to do everything. You'll get a good overview of the entire operation. You'll get hands-on experience in shooting and editing. At large market stations, you won't be allowed to touch the equipment because of union restrictions. The trend at the larger market stations is for workers with talent in a variety of areas. So, if you have edited, shot, written and reported your own stories, your prospects for landing one of these hybrid jobs of the '90s are greater.

However, Professor Hewitt warns about a common practice in small markets involving on-air reporters and anchors. "Small market people know they're being used. They'll often ask you to sign a contract for one or two years. Be aware. Sometimes there will be a top-10 or top-20 out, meaning you can leave if you're offered a job in a top market. Sometimes they get very vindictive and hound you if you break the contract, or sue you. It is legally binding. If they want a contract, try to negotiate for a one year length with a top 40 out." He also warns, "Don't go to a market where you don't think you'd want to live. Sometimes a small town can be a tough adjustment, especially if it's in a part of the country where you've never lived."

Wendy Burch, weekend anchor and reporter at WXIX, the Fox affiliate in Cincinnati, agrees, "Target the areas you know you'd like to work in. Know where it is you can be happy, because being happy is just as important as getting your foot in the door." During her junior and senior years at Brigham Young University, Burch landed a coveted part-time job as desk assistant at KSL in Salt Lake City, the 41st market. She says, "I was a very small fish in a huge pond. I realized no one was going to discover me, and that I had to go outside."

In 1990, as her graduation approached, she "shotgunned tapes everywhere, whenever there was a job opening." After a few months, she got an anchoring job at KIEM in Eureka, California. She said, "Going to market 185 was the greatest experience of my life. I walked into that market and was the 6 o'clock anchor, the main woman in Eureka. It was great. I had a wonderful time. I really did do it all. The first day I walked in, they said, 'Here's the weather board. This is where you'll be doing weather.' I said, 'Scuse me??!!' I did weather. I produced the show. I was the news director. I was the reporter. I acted as assignment editor. As executive producer. And worked out budgets, hired new people, and in my spare time, swept up and did windows as well. It really prepared me for moving up to larger stations."

She advises, "Don't be too choosy. Don't look at the market size. Look at the opportunity. Take road trips [travel from market to market for informational interviews—see chapter 13 for more on this]. It gives new meaning to 'we leave the light on,' because you know you're stopping at every Motel 6 along the way. That's the only place you can afford. But that's a great way to know what you're getting yourself into. I went from working for a wonderful station in Salt Lake City with all the bells and whistles and toys, to a trailer in Eureka, California, next to a cow pasture. It was a shock, but I knew what to expect, so I didn't break out in tears when I faced my new working conditions."

She warns, "Know that a salary in a triple digit market in this economy is not good. Be prepared to make less than $15,000 a year. I wasn't, because I entered with a million and one student loans and a Mastercard bill up to the sky. So prepare yourself. You'll have moving expenses and the station might help you, but not always in a smaller market. Be ready to dress and look the part. You have to get out of that college mode of jeans and a T-shirt and look at how to project. It's really important. And remember it in your resume tape. You're selling your image. Be prepared to leave your friends and family and that comfort zone. And that includes boyfriends and girlfriends. That can be difficult because you often just have a couple of weeks to get there. And know that there is a lot of competition out there. Two hundred people applied for my position as the anchor in Eureka. And if you work just as hard to get out of that market as you did to get into that market, you can find yourself somewhere bigger and better." She did just that, moving to Reno, the 117th market, as a reporter and morning anchor after only 10 months in Eureka, and then, 21 months later, on to Cincinnati, the 29th market.

Bill Groody says, "As a strategy, it's better to go to a smaller market as a reporter than to go to a larger market as a desk assistant, hoping you will be discovered and turned into a reporter. People want to see skills, they want to see that you can handle a job. You can't put together a resume tape being a desk assistant someplace." Groody says a lot of talented people landed desk jobs at NBC right out of college and got very comfortable making a good salary, but not doing what they wanted to do. He said, "People would homestead in those positions, and in the past five or six years, the number of desk assistants was cut back drastically, and all of a sudden they're in their 30s and 40s and they're stuck asking themselves, 'What do I do?'"

If you're not willing or able to relocate, you simply limit your options. When your marketplace encompasses 200 markets instead of just one, you can see how much greater your chances are of getting in and moving up.

See Appendix A for all the markets in the U.S. of A. and their relative size.

Take Any Job or Hold Out for the One You Want?

This is always a big area of debate. At career panels over the years, I heard the human resources folks advise job seekers not to take a job in an area that didn't reflect their ultimate goal. In other words, don't take a job in traffic or as an administrative assistant in programming if you want to work in news. You'll also find many department heads trying to weed out applicants for secretarial positions who are highly motivated for other positions. One of the reasons is that once in, these bright enthusiastic newcomers often become very impatient to move on and become discouraged if they aren't promoted within a few months. They bug, heckle, needle, and nag their supervisors unmercifully until said supervisor is swearing that the next person hired for the job will be someone who actually wants *that* job.

On the other hand, a TV station's entry level positions are often filled with la creme de la creme—energetic, sharp, hardworking recruits who will do everything and anything to prove they're capable of doing much more and that they're worthy of a promotion. In fact, TV stations can usually find a few nonpaid interns who, after just a couple of weeks learning the ropes, are doing the work of production assistants, publicists and administrative assistants, and contributing to the station's profits.

Having a great staff of administrative assistants, production assistants and interns does come in handy when someone is needed to fill an abruptly vacated position, either permanently or during a temporary leave of absence. Or a bright administrative assistant might have her job redefined to incorporate some of the responsibilities of another job when a station is in the process of downsizing.

From the individual's point of view, there are definitely pros and cons to taking any job you can get. First, the pros. Once your foot is in the door, you do have a much greater opportunity to meet people, both in the station and at other stations, via the phone or at industry events or conventions. And having those call letters on your resume will definitely help you get that next job at another station. This is particularly helpful if you've got a big market station's call letters on your resume and you're applying in a smaller market for a bigger job. The smaller market folks may even hire you to take advantage of your contacts at the bigger station in their attempt to move up to a larger market.

On the negative side, you could get stereotyped as a sales assistant or administrative assistant and have a tough time breaking out. Or, as I mentioned earlier, your supervisor may not want to let you go because you're so valuable.

Departments vary in how much mobility they afford. Generally speaking, the more creative the department, the more mobility. The sales, traffic, accounting and station administrative (corporate) departments can be dead ends, whereas creative services and news are at the other end of the spectrum.

There's more movement in a smaller market station due to a lesser degree of specialization and lack of unions. Employees get much more breadth of exposure to various jobs in the smaller station. In most of the larger markets, the technical jobs, and sometimes the newswriters, reporters and anchors, are union positions, making it much more difficult for the new kid on the block to learn or move about. Duane Fulk, who's been an editor in the business for more than 15 years, warns that the technician jobs are traps if your interest is in producing, writing or reporting. He advises against accepting jobs as video editors, shooters and engineers, unless that's your ultimate goal.

One more factor in advancement is the attitude of your specific supervisor. Some department heads really are wonderful about developing the potential of their staff, either promoting them or encouraging them to pursue other opportunities for advancement. My first TV boss was just such a saint. When Dennis Fitch hired me as a secretary/publicist at KGO, I let him know I definitely had aspirations to move beyond my secretarial responsibilities ASAP. He assured me I'd be promoted within a few months, and made good on his promise. Other bosses, for whatever reason, be it not wanting to find and train replacements, or because they fear competition or have chauvinistic attitudes, hold their employees back.

I believe that being honest with a potential employer is the best policy because that person can become your best ally. At the same time, assure him or her that you will work your butt off in the job you're hired to do and you will be eternally grateful for the opportunity to do so.

In a gross oversimplification, my advice is to take whatever job you can get, and after you're in, be proactive in learning all you can and making contacts *while doing the job you were hired to do,* cheerfully and with excellence.

10 THINGS TO CONSIDER BEFORE ACCEPTING A JOB

- *Are there opportunities for advancement?*
- *What will I be able to learn?*
- *Is this a company where I would like to work?*
- *Will I be happy living in this town/city?*
- *Can I afford to take this job, or find other ways to supplement my income?*
- *Is my supervisor someone who will support my goals?*
- *Is this a good place to find a mentor?*
- *Will this experience enhance my resume?*
- *Do I get a parking space?*
- *Are there other perks of the job?*

Diversity: Making the Most of Yours

What about racism, sexism, ageism? If you're spending a lot of time on this question, you're in trouble. Prejudice and discrimination exist to varying degrees in different markets, different stations, different departments and different hearts and minds. If your goal is to change peoples' attitudes, passed down for generations, you're talking about a full-time career. Besides, the best way to change their minds is to prove them wrong—by your actions, not your words.

The only person you really have any control over is yourself. And the more time you spend worrying about others' prejudices, the more you're undermining your own confidence and letting your negative thinking sabotage your chances. I've said it before, and I'll say it again: There is a self-fulfilling prophecy here. You have to believe in yourself and your abilities. The key is discovering your talents and your passion, and focusing on these.

Times have changed. I remember the '50s well, when you never saw an African-American on TV—not on the newscasts, in the programs or even in the commercials. The only women on newscasts were the occasional "weathergirls." Today, stations, particularly in the larger markets, have diversified in all departments, with the exception of upper management, which is still predominately white males. There are a number of women in what used to be the totally male-dominated areas of engineering and operations. On the air, woman are in the anchor seats, in the field, they're reporting news *and* sports, and behind the scenes, they're producing, writing and directing.

Today, you're just as likely to hear white males crying "unfair." They're concerned that less qualified women and persons of color might be getting "their" jobs. It's extremely competitive for everyone, and if it seems infinitely more competitive for men today, it's because the competition is tougher. Not only are more and more graduates with broadcasting and liberal arts degrees attempting to break in, but entire groups—women, minorities—are no longer being denied access.

If you are a woman or minority, there are associations that can be of tremendous help. Some are national with local chapters, and others serve individual markets. In Northern California, many have benefited from the Bay Area Broadcast Skills Bank's services. The ones in your area will be easy to locate with a little research and networking. (See chapter 11, Networking, for specific associations.) The first ever Unity Convention was held in Atlanta in the summer of 1994, bringing together various groups of journalists, including the National Association of Black Journalists (NABJ), the National Association of Hispanic Journalists (NAHJ) and the Asian American Journalists Association (AAJA). It offered great educational and networking opportunities. In the cable industry, the Walter Kaitz Foundation was founded specifically to recruit ethnic minorities for management, professional and technical positions from outside the industry. (See chapter 24 for more on the Kaitz Foundation.)

Also, some TV stations, networks and production companies have minority trainee positions, some of which are open to women as well. These positions, which are few and far between, usually last a year and most pay minimum wage, but they are definite stepping-stones to careers in broadcasting. Janine Knudsen of KRON's human resources department told attendees at a station career panel, "After the traineeship, you're not promised a position, but you hear about things as you go along. All of the trainees here have been offered jobs at the end of their year. Only one did not take it, for personal reasons. One of our recent trainees worked at KRON six months after her internship ended and then got a job with 'Oprah'."

In 1979, after graduating from Stanford University in communications and political science, Javier Valencia landed KRON's minority trainee position. Two months into the year-long program, he was offered an entry level job in the station's public affairs department. Today, as KRON's community relations manager, he says, "I benefited from community activists who worked hard to convince stations that they needed to reflect the diversity of their communities, both on the air and in the work force. I was in the right place at the right time. I think many minorities currently working in broadcasting, myself included, are committed to mentoring the next generation."

SLEEPING YOUR WAY TO THE TOP

Sorry, this really isn't a viable way to get into the business or advance once you're in. The business has gotten way too competitive to allow room for lightweights who can't do the job. And there are laws, being enforced at this writing, that prohibit hirers from demanding sexual favors in return for job offers. That said, I'll never forget being introduced to Betty Hudson, NBC's first female senior vice president. She told me that a good male friend at NBC gave her the perfect response for any man who had the gall to ask, "Who'd you have to sleep with to get to where you are?" She says, "I just tell them, 'Everybody but you!'"

Janette Gitler asked Fred DeCordova, then producer of "The Tonight Show," if it was true that in Hollywood people have to sleep their way into a job, or is there something about being attractive that helps one get in? Janette said, "He was very honest about it. He said he couldn't deny that in some cases some people might sleep their way into a job, but there was no way they were going to keep it unless they could do it well. He said that if he was hiring a talent coordinator and had the choice between two equally qualified people, it is likely that the more attractive person would end up getting the job. He explained that the job of a talent coordinator involves going out to lunch with celebrities and stars, and someone would much rather go out to lunch with somebody who's attractive than with somebody who isn't attractive."

It is fair to say that good looks are an asset for both men and women in TV, and frankly, not just for on-air positions. And not just in TV. To illustrate the normal human bias towards good-looking people, a network news magazine staged two identical trials, with two different juries who thought the trial was for real. The lawyers, witnesses and judge were "played" by the same people in both trials. Only the defendant was different. In one case, a good-looking man was on trial. The other defendant was average-looking. Neither defendant took the stand. Deliberating over identical evidence and testimony, the jurors in the first trial found the good-looking man innocent. The other set of jurors found the second man, the average-looking defendant, guilty. When interviewed later about their verdicts, jurors attributed all kinds of positive attributes to the attractive man—he seemed honest, sincere, etc. The second set of jurors had made negative assumptions about the less attractive man. Neither man had spoken during the entire trial.

So, looks do make a difference. It may not be fair, but it seems to be human nature. And I've known hirers who have been unduly influenced by looks, but the pressure today is so great on making more money with fewer and fewer people that there's no room at all for fluff or window dressing. Besides, there's not much you can do about it except look your best.

THE QUESTION OF AGE

Which brings us to the question of age. This is frankly a young business. Hirers in the world of television are looking for applicants who possess unbridled enthusiasm and are eager to learn, excited about working on the latest equipment and computers, willing to put in long hours, and happy to do anything—for next to nothing.

Young people, who are more likely to put up with several roommates to share expenses or take advantage of Mom and Dad's hospitality, can often better afford to start out for minimum wage in a TV job. A person with a lot of flexibility, without children, able and willing to move from market to market, has the edge breaking into the business.

It's very difficult to get in without starting at the bottom. A number of people have made the transition from radio to TV, others from print journalism to TV, but even other media experience doesn't assure an entrée.

I've met a number of people who wanted to break into the business in their late thirties and forties, but very few of them do. Still, don't let statistics get in your way. You could be the exception to the rule. And you won't know if you don't try.

Deborah Guardian, assistant to the television station manager at KQED, San Francisco's PBS station, broke into the business at age 35. Not only that, but she landed the job by sending a cover letter and resume in response to a classified ad in the local paper! Her background included international publicity for United Artists in New York, hotel management, human resources director for a hospice and PR for a nightclub. In her early 20s, as director Milos Forman's assistant, she spent an entire year in Czechoslovakia on location with the feature film, *Amadeus*. She was accustomed to working in a fast-paced, high-energy environment, and doesn't have a 9 to 5 mindset. No doubt her attitude and varied background gave her an edge.

The bottom line, according to Don Fitzpatrick of Don Fitzpatrick & Associates, the country's leading on-air headhunting firm, is, "If you're going to be hired in the '90s, you've got to be better than the people sitting around you. Which means, you've got to market yourself. You've got to have the best resume, you've got to have the best resume tape, you've got to have the most crisp letter that will attract the attention of a news director or a general manager." Although he was speaking specifically about on-air talent, his advice applies to all areas of broadcasting.

Tools of the Trade

Resumes and Cover Letters

Your resume may be the most important document you develop in your life. Your career could depend on it. Think of it as an important tool in the toolbox that will help you build your career. A masterfully constructed resume will not only reflect who you are but will add to your confidence during your job search, whether you're networking on the phone or in the interview for a real job.

You want to get your resume crafted, sanded and polished before you do anything else because you always want to have a few with you. You never know where or when you'll meet someone who might have a lead for you, or even a job. Send out your resumes with equally well-crafted cover letters to prospective employers or to contacts who can put you in touch with prospective employers. These two signature pieces are usually your first introduction to a busy professional, so they must be compelling and succinct.

RESUME BASICS

Here are some very basic rules for the resume, agreed on by everyone I've met in the business. Many of these apply no matter what industry you're pursuing.

- ***Confine it to one page.***
  Although you may feel that everything you've done is important, the person reading your resume is very busy and wants an overview. If more detail is needed, you can elaborate on your qualifications and past experience over the phone or in an interview.
- ***Use action verbs in descriptions.***
  Managed, coordinated, developed, initiated, created, wrote and edited, instead of "responsible for" or "duties included."
- ***Perfect it—no typos, misspelled words or incorrect grammar.***
  I used to sort resumes into two piles—those with errors and those without. This usually cut my employee search time in half. I simply didn't consider anyone whose resume contained a misspelled word or typo. My reasoning was: If you're sloppy with your own resume, upon which your very livelihood depends, how conscientious would you be with your facts and detail orientation on the job? Even less accurate, one would expect, especially when you factor in deadline pressure. In the TV business, you may be dispensing information to lots of people, either over the airwaves, or if you're in PR, via press releases. Accuracy is essential. Mistakes are fruitful and multiply. They also reflect poorly on the boss. In fact, the boss takes the rap. This makes the boss very unhappy.
- ***Print it professionally—on a laser printer on premium paper.***
  With such stiff competition, your resume must be more than legible. It must reflect professionalism. If it's not in a class by itself, it should at least be at the head of the class. With computers and laser printers, a resume typed on a typewriter of yesteryear, or a resume with a glob or two of white-out, stands out like a sore thumb.

- ***Put your education at the bottom of the page, not at the top.*** A common mistake by a novice with little experience is to list education credentials at the top of the resume. This format simply screams, "I have no experience that I think is of value." It's a red flag for a prospective employer.
- ***Don't include irrelevant information, even if you have some space to fill.***
  Hobbies, marital status and grade point average are not only unnecessary, but detract from the professionalism of your resume. Perhaps if there's something that might help you stand out, you might put it in, but first test it out on a couple of friends in the business to get their reactions. On my resume, in addition to B.A. Sociology, Smith College, Northampton, MA, 1969, I added Chateau Mont Choisi, Lausanne, Switzerland, 1962. It's not necessary to list that I'd studied abroad for a year during high school, but it does indicate that I'm bilingual. Although speaking French was irrelevant to the positions I was seeking, perhaps it separated me slightly from the competition and resulted in an interview or two I might not have gotten otherwise.
- ***Tell the truth.***
  Lying on your resume will come back to haunt you. It's not one of those qualities folks in journalism are seeking. Stretching the truth is something they get plenty of from politicians, lawyers and corporate spokespersons. However, as a former public relations person, I would be remiss if I didn't tell you that *how* you tell the truth can make you look good, bad or indifferent. Just be sure that what you say in your resume or interview will stand up in the court of closer scrutiny.

In addition to the above resume basics, there are a number of other issues to consider which are somewhat more subjective. Some interviewers want job descriptions on the resume, others think that the title of the position tells the story and a description isn't necessary. Some people like a job objective at the top, others feel it's irrelevant and limiting. Some say you should list every job you've had, others agree you can be a bit more selective. Some like references listed because they want to check you out and don't want to have to follow up to your line, "References on Request." Others believe any references you'd list would be unduly biased (or you wouldn't have listed them). If they're interested in you, they'll find the people you *didn't* list on your resume to call.

Professor Hewitt has some sound resume advice for those looking for jobs in news: "Everyone in news knows what news people do. You can't BS on your resume. Your resume should reflect your experience, what you've really done. It's very important how you write the description of your experience. Say what you did. If you worked as an intern on a news assignment desk, say, 'made phone calls, monitored police scanners, etc.' Listing your actual duties will reinforce your confidence. At the end, say, 'as an intern.' Do not leave it off but don't highlight it by putting it at the top. Make your resume sound professional. Just because you were an intern doesn't mean you weren't doing the job. And don't overlook projects you did in school—as interviewer, producer."

You want to focus on the experience you have that is most applicable to the job you're seeking. If you've had only a little TV experience, for example, as an intern, you'll want to make the most of that if you're applying for TV positions, and downplay your experience as a salesperson at The Gap or hamburger flipper at McDonald's.

Of course, if you've had very little job experience, you may have to go into a bit more detail about job experiences just to fill up the page. But don't be afraid of a little white space. Your interviewer will appreciate a straightforward, easy-to-read resume.

In the '90s, as mentioned earlier, the TV business is looking for people who can perform a variety of functions. Find a way to express in your resume your innate versatility.

Some recommend you redo your resume for each job you seek. That may be relatively easy if you have your own computer and good printer. If you don't customize your resume, I recommend deleting a job objective. Stating a job objective can turn the tide against you if the job that is open isn't in line with your objective. An interviewer might not tell you about other opportunities in the programming department down the hall if you've stated you're seeking a career in news. A career path can be a long and winding road, and it's better to be on the path than not on your way at all, sidelined in your search for job perfection.

Rosemary Wesela warns of a common error she's seen over the years, "If you're serious about an on-air career in journalism, don't put on your resume that you want to be an anchor or a host, because they don't go hand in hand. Non-news TV programs are completely different from programs that originate in the news department. Journalists are offended by people who say, 'Well, I either want to be a journalist or I'd like to host that show that you have on Saturday morning.' Have different resumes and different resume tapes if you're willing to go either direction. In a big market, you quite frankly get laughed at if you put them on the same tape or resume."

THE RESUME FOR THE PERSON WITH NO RELATED EXPERIENCE

When Darryl Compton was making most of the hiring decisions in KRON's newsroom (including for all of the entry level positions), he used to tell the career panel guests he just scanned the resumes for station call letters. Where a person had worked previously was his primary concern. This is why internships are important. They'll give you both experience and call letters for your resume.

The less experience you have in the business you want to break into, the better job you need to do on your resume.

When I decided to take a stab at a TV career in 1974, I signed up for a weekend seminar called "Career Design" offered by Ranny Riley & Associates in San Francisco. The curriculum's goals were threefold: (1) to help participants discover their aptitudes and interests and help them target the fields they wanted to go after, (2) to compose the ultimate resume drawing on work, volunteer and educational experience, and (3) to map out a strategy of informational interviewing and networking to discover and take advantage of the "hidden job market," jobs not listed in the Sunday classified section of the local paper.

To make a short story even shorter, I compiled my resume during the weekend course, had 300 copies professionally printed and scheduled a week's vacation to conduct informational interviews. On the second day of my interviewing, I interviewed for a wonderful job at KGO, the ABC owned-and-operated TV station in San Francisco. A couple of weeks later, that job was mine, and it turned into a 20-year career, a lot of fun and a very good living. (See the end of this chapter for the resume that got me in.)

If you're trying to break into television, or any new industry, you're faced with the challenge of making your skills look good enough on paper to bridge the chasm of inexperience. You want to emphasize your transferable skills, instead of the duties you performed that are specific to some other industry.

One of the techniques I learned from Ranny Riley was to separate jobs from skills. You list your job experience under a subhead of "experience" at the top of the resume—titles, company names and years worked. I called myself assistant to the creative director and assistant to the promotion director rather than secretary, as I was determined to break out of the secretarial rut. Under a second subhead of "areas of effectiveness," you list accomplishments that emphasize your skills. You can group these accomplishments under skill areas such as writing, program planning, budget management. This format allowed me to omit the bulk of my experience, which was typing, coffee service and phone call placement and reception.

Dennis Fitch, the promotion director at KGO who gave me that first job, was looking for someone who could write press releases, program listings for *TV Guide* and an occasional feature article for the entertainment section of the *San Francisco Chronicle*. My past work experience included very little writing—just a couple of travel brochures. But when he asked during the interview if I could write, I said, "Yes, I wrote this resume, for example." To which he replied, "My God, if you can promote yourself to this degree, I guess you can promote KGO!" Just as curiosity and digging for a story are attributes appreciated by those hiring for news jobs, a measure of creativity, which allows you to make a silk purse of a resume out of a sow's ear of experience, is appreciated by promotion directors.

Another benefit of listing areas of effectiveness is that you can include accomplishments for which you weren't paid. Half of the items I listed under areas of effectiveness were not part of my paid experience. They were either assignments I took on at college or as a volunteer outside of work, for nonprofits or professional organizations, or as a volunteer within my company. For example, I coordinated funny slide presentations, taking photos of employees and working on the scripts, for the annual company Christmas party—an internal affair, but it was experience nonetheless.

Hirers weigh resumes carefully, comparing experience and background, but they also pay close attention to their gut feeling. Who would be the best addition to the department, be self-starting, a quick study, flexible, accurate and have a terrific attitude? It often isn't the person with the most experience.

For someone like myself, hiring an employee is like buying a house. You make the decision that you have to have this employee or that house in the first few seconds. Then you spend the rest of the time justifying your decision. So in your resume, you want to give the hirer something on which to hang their hat of justification. Because if you don't work out, most employers want to at least be able to say, "How could I know? She had this great resume." By all means, make your resume your ally. Don't let it be your worst enemy.

WHEN TO SEND A RESUME

Soledad O'Brien was a senior at Harvard majoring in English and American Literature just a few years ago when she landed an internship at Boston's WBZ in their "made-for-TV movies" department. One errand to their newsroom was all it took to get her hooked on TV news as a career goal. Upon graduating in 1989, she sent out 178 resumes with cover letters to news directors in the ten cities she targeted. In her cover letter, she said she would take any entry level job.

Soledad eventually received 30 responses from the 178 inquiries. All but three of the 30 were form postcards, stating that the station would keep her resume on file. Three were nice letters, but only one led to an interview and that was at WBZ, where she'd interned. The news director there forwarded her resume to the human resources director and she did get an interview out of it for the minority trainee writer position. (That program has since been discontinued.) She didn't get that job, but she did get hired as the assistant to WBZ's medical reporter for $7 an hour.

The point is that mailing out resumes and cover letters that aren't targeted to specific job openings is a shotgun approach. Don't be discouraged if you don't get positive responses or even any responses beyond the form postcard. If a news director or another department head receives your cover letter and resume and has no openings at the time, it won't get more than a cursory glance on the way to the circular file or the human resources' processor of postcards.

Many human resources directors suggest you send a cover letter and resume *only* in response to a specific job posting. If two jobs are posted for which you'd like to apply, you should send a cover letter and resume for each position. Frankly, TV stations may say they'll keep your resume on file, but they don't say anything about looking in the file. When jobs open up, they rarely take the time to look back for qualified applicants, because who knows if those applicants are still available or interested? Usually, they can choose from plenty of fresh, new applicants responding to that specific job posting.

Don't worry about sending cover letters and resumes to the same TV station over and over. Eventually, they may remember your name and feel as if they know you. And, who knows, you may even be invited in to interview for a job.

You want to send resumes and cover letters in response to job postings, to set up appointments for an upcoming road trip, and to let department heads in targeted cities know you're available.

For your mailing to be effective, follow up on the phone.

COVER LETTERS

Good writing is a cherished commodity in almost any non-technical TV job, and your cover letter will be your first opportunity to show off those writing skills. It's also your chance to show that you have other desired qualities—a good attitude, enthusiasm, intelligence and dependability.

Never send a resume without a cover letter. A resume traveling solo elicits the response, "So what?" The cover letter answers that question.

Fred Zehnder says, "The letter should be short and to the point. It's essential that the person have a clear idea of the job they want. I've gotten letters that have said, 'I am interested in getting a creative job in the media.' That person has no idea what they want to do. They don't even know what the media is all about. It'll be a waste of time to talk to someone like that. If the letter says, 'I'm really interested in the assignment desk because I like to make decisions about news stories and listen to the police radio,' then I'm much more inclined to meet them."

Rosemary Wesela, who has had hundreds of cover letters cross her desk in the KRON newsroom, advises, "The letter should say everything you would want to say if you were sitting in the room, but you have to edit it down to one page. It's important that the news director see that you can be concise, because if you're working on a story you've been covering for four hours, you're going to need to tell that story in a minute-thirty. Always consider that you're selling yourself to a news director. You should use the same skills you would use as a reporter."

Your cover letter is your chance to focus on the particular job you're applying for. An innovative letter can stand out and help you rise above the crowd.

SAMPLE RESUME

This was the resume that helped me land my first job in TV. Note how it accentuates the positive. Few would guess that four-fifths of my work experience was secretarial. At least a third of the activities listed under "areas of effectiveness" were volunteer assignments. In fact, the "program director" position under the "experience" heading was also unpaid.

LINDA GUESS	Address City, State ZIP phone number message or fax number
Job objective	A position using demonstrated skills in communications, program development and coordination. Special interest in media, tourism and women's resource development.
Summary of experience	*Program Director*, 1974 San Francisco Women in Advertising, San Francisco *Assistant to the Creative Director*, 1972-present *Assistant to the Promotion Director*, 1970-72 Clinton E. Frank, Inc. Advertising, San Francisco *Travel Coordinator and Tour Escort*, 1966-70 Nob Hill Travel Service, Inc., San Francisco International Travel Agency, Inc., Houston
Areas of effectiveness	*Coordinator* Coordinated activities of creative department of advertising agency including meetings, film screenings, audio/visual presentations; served as liaison between creative director and department of 15. Supervised distribution of company newsletter to 1,600 new business prospects, clients and suppliers. Directed survey of professional women in advertising; designed questionnaire, tabulated and evaluated the results for publication. *Travel/promotion* Created and implemented group and individual itineraries; handled domestic and international travel arrangements for commercial accounts. Escorted tours to Hawaii and the Orient for groups of 80 to 100; served as cross-cultural liaison, mediator and problem-solver. Initiated and implemented contests, parties, travelogues, tie-in promotions; handled all arrangements for winners of trips. *Communicator* Edited educational programs designed for elementary and secondary schools on the functional organization and operation of various industries; distributed programs to 300 schools. Coordinated slide presentations; served as photographer; worked on scripts. *Program developer* Co-founded group of professional women in advertising; planned and developed monthly programs, recruited speakers and conducted preliminary orientation meetings with speakers and panel members. Developed orientation program for college freshmen to facilitate their adjustment, assist them in selecting fields of study, counsel them and provide information about facilities and opportunities.
Education	B.A. Sociology/Economics, 1969, Smith College, Northampton, MA French Degree, 1962, Chateau Mont Choisi, Lausanne, Switzerland

Networking: Resources, Contacts and Associations

WHO DO YOU KNOW AND WHO DO THEY KNOW?

You've heard it a million times: "It's not what you know. It's who you know." There's some truth to it, although as I said earlier, once you get that interview, you've got to make the grade on your own.

Professionals do not like to be pressured into showing favoritism to friends of the owner, but it does happen. In my years in local television, I was occasionally "assigned" interns. They were often from well-to-do families, and they probably wouldn't have been chosen for the internship if they had competed in the open market because they usually lacked that hunger.

Most aspiring broadcasters don't know the owner of a station, so most of the networking takes place at a lower level, which is, in fact, preferable. A news director is happy to talk with a friend of a valued producer or colleague, and do a favor for a friend. But when the pressure comes from on high, the news director may feel manipulated and compromised.

After graduate school, Joe Fragola contacted a close friend of his family, Bob Keeshan. Here's how Joe tells it: "Bob was at CBS then, still doing 'Captain Kangaroo.' So I asked if I could just meet with him. I wanted advice. He said sure. I went in, caught the end of a taping, sat down with him and he asked me what I wanted to do. I wanted to be a producer. I knew I wanted to have creative control. I knew I wanted to work on a project. I liked coming up with an idea, following through. And telling good stories, and maybe sometimes moving people, and having a different ability to sometimes tell stories in a different way that was maybe my own signature. So I told him that, and I said, 'But I have to start at the bottom, because when I get to the top I don't want anyone to say I didn't earn my stripes, that I didn't pay my dues.' He was talking to my dad the next day. He said I was the first kid who'd come to him since 1947 when he started out as Claribelle on 'The Howdy Doody Show' who didn't want a $50,000 a year job when he got out of school. He thought that was very refreshing."

Joe landed his first job in 1972—as a CBS News production assistant on Walter Cronkite's newscast. Joe says, "Bob told me to call personnel and go down and talk to them. He may have used some influence, but the main thing is, I had to then go down and talk to the bureau chief for New York, which is the network assignment desk person that covered for 'CBS Evening News.' He looked at me and my resume with my Master's degree in mass communications and said, 'You're better educated than Walter.' I said, 'I know, but give me a broom, I gotta start some place.' Maybe it was the right day. Maybe he didn't have a hangover or something. But he said, 'okay' and he hired me."

Duane Fulk, now a freelance editor, got his start through a friend of a friend running cameras at race tracks, first at the Alameda County Fair, then at Golden Gate Fields and Bay

Meadows. It was seasonal work. (He brags that he followed the horses for two years and never bet on a race.) He says, "Although it was a closed-circuit operation, it gave me some experience for my resume and it got me into the union. The same friend who'd been my contact at the racetrack got a job as a photographer at Channel 36 in San Jose. They needed an editor. Somewhere along the line, I learned about editing. My job there was cutting time out of the feature movies to fit in the allotted time slots. It was a great learning experience for someone who wanted to make films, which is what I wanted to do when I got out of college. You'd learn how these films were put together by taking them apart."

Your contacts don't have to be close personal friends. Libby Moore got her job as personal assistant to Maury Povich through a connection. She explains, "My younger sister's best friend's mother's good friend was an associate producer on Maury's show. I met my sister's friend's mother once, but I'd never met her friend. I wanted to write for 'Saturday Night Live,' but I called this producer who said, 'Get the hell out of Boulder, move to L.A. or New York and call me when you get here.' I did, and I told her 'I don't care what opens up, I'll do anything'."

You start with who you know. And who they know. Start with your family. Do any of your relatives know anyone in the business, or anyone who knows anyone? And you work your way through all your acquaintances—from school, work, former employers, neighbors, clubs and social or civic organizations, merchants, people you met while traveling. Get on the phone and talk to everyone you know.

Your best contacts will be the ones you develop through your own networking, because you can target who you meet. Join professional organizations, go to conventions, find out where people in the business are speaking. If you're interested in news, find out from a news director's secretary if s/he'll be speaking to any community groups and attend those functions if at all possible. Many TV professionals teach seminars or courses at community colleges. That's a great way to develop a relationship. You can also use the technique I mentioned in chapter 5 about finagling a tour of a station and meeting some people along the way.

News Director Doug McKnight of KCCN in Monterey, California, confirms the notion that it doesn't hurt to have some contacts in the bank. He says, "You want to get to know me when I don't have a job open, so I'll know who you are when I do."

Your local chapters of the National Academy of Television Arts and Sciences (NATAS), RTNDA and AWRT all have speakers, panels, seminars and social events. All of these afford major networking opportunities. I'll elaborate in detail later in this chapter when I talk about associations.

RESOURCES

You want to approach getting a job in TV like you would getting a story. You go out and do the research and find out where everybody is, who everybody is, what stations are the best ones to work for and which ones have the openings. Your research should start with the easily identifiable resources at your fingertips.

First, if you're in college, take advantage of your broadcasting department's resource listings and the school's career counselors.

Next, visit your library in person or via your modem. Use the *Broadcasting & Cable Yearbook* to learn more about the stations in the markets that interest you. Read the trade journals to familiarize yourself with what's going on in the marketplace. The most informative are *Electronic Media*, *Broadcasting & Cable* magazine, *Variety* (daily and weekly), *Hollywood Reporter* (daily), *Multichannel News* and *Cablevision*. Even *AdWeek* and *Advertising Age* cover TV. Several of the publications have job listings for national and local TV positions in their classified section, occasionally even for entry level jobs.

Develop a list of job lines. Many stations have recorded listings of job postings. Most are updated only when a new job opens up; however, you should call once a week because often jobs are only posted for a couple of weeks. After that length of time, a station may have a couple of hundred resumes and may not accept any more. It's important to respond promptly when you do hear of a job opening to be sure your resume arrives prior to the deadline.

In addition to individual stations, there are other job lines of professional organizations and associations. RTNDA has a national job line with daily updates of listings (1-900-40-RTNDA, 85¢ per minute). Their members also receive a printed biweekly job bulletin at no charge. Many organizations such as the California Chicano News Media Association in L.A., the AAJA and NABJ have job lines for their members. And in larger markets, you can easily find the names of media groups that serve the general public or special groups.

ASSOCIATIONS: THEIR CONVENTIONS AND MEETINGS

The ultimate in networking is hanging out with people in the business, attending their meetings and conventions, becoming a fly on their wall, then a bug in their ear.

Wendy Burch, the anchor/reporter in Cincinnati who got there in just three years via Eureka and Reno, reports that when she was starting out, she saved her money to go to the national RTNDA convention in Denver. She says, "I was able to speak with and meet news directors face to face who I couldn't get to on the phone." In addition to news directors from radio and TV stations all over, RTNDA attracts producers, newswriters, consultants, and some students and go-getters working to break into the business.

If news is your game, RTNDA is the place to see and be seen. True, news directors will have their own agendas at the conventions and won't want to spend an hour chatting with you in one of the hospitality suites. But in a few minutes, you can ask them a couple of questions, make an impression, perhaps give them a resume, and ask if you can call them when they get back to their offices. If they're on panels, you can ask intelligent questions related to your job search, prefaced by your name, your degree from what school and a little about your experience. Later, if you don't get a chance to say hello in person and deepen your budding relationship with a couple other questions, you can still call them, tell them how much you appreciated what they had to say on the panel and remind them that you were the one who asked that great question.

Don't expect everyone you meet to greet you with open arms. Fred Zehnder is not too encouraging about meeting upstarts at RTNDA. He says, "You have access to about 2,000 news directors. It's an opportunity, to be sure. But to be honest with you, it bugs me. It doesn't work with me. But it may work with news directors from smaller stations. I go to the sessions and want to see other colleagues. I do, however, bring resumes back with me."

Over and above meeting valuable contacts, RTNDA and other conventions offer great speakers and seminars about the industry and its future. RTNDA's convention in Los Angeles in October 1994 had sessions on "News of the Future," "Getting Started in Computer-Assisted Reporting," "Living in a 500-Channel Universe" and what that will mean for local news, and "The Urban Crisis: Covering Crime and Health in America's Streets." Other sessions focused on environmental reporting, political reporting, how court decisions affect what goes on the air, and newswriting skills. Some of the better known speakers and moderators were Charles Kuralt, Jane Pauley, George McGovern, Andy Rooney, Cokie Roberts and John Sununu.

Some sessions are geared to students. One such session focused on making the most of your school's resources to turn out a good student newscast. Another zeroed in on getting an internship, and another on looking and sounding like a pro. Students are also invited to have their resume tapes critiqued.

Similar sessions for reporters and anchors already in the business offer tape critiques by a news director, voice coach, makeup/hair/clothing expert and news consultant. The RTNDA regional conferences also offer tape clinics for professionals.

Recent Northern California RTNDA regional conferences have included sessions on getting your foot in the door, and also "A Foot Out the Door," or what to do when you've been downsized.

Other sessions have included "Sex in the Newsroom" and "Cross-Training: The Superjournalist."

At national and regional meetings, you can post your resume or consult the posted job listings on a job bulletin board.

The national conference features a student rate of $65 which includes all events, but meals must be purchased separately. For regional RTNDA meetings, corporations sometimes make contributions to permit students to attend at a reduced fee. Call and ask about scholarships.

There are benefits to joining RTNDA, including continuing education, a biweekly job bulletin, and a monthly magazine, *The Communicator*, covering industry news, trends and technology.

For more information:
Radio-Television News Directors Association
1000 Connecticut Ave., Suite 615
Washington, DC 20036
(202) 659-6510

If you're interested in other areas of broadcasting, programming, promotion and engineering each has its own convention. Call to find out about locations, special rates and the subjects they'll be covering. Also ask if they have special sessions for students or others wanting to get their start in the business.

For programming, look into the NATPE convention, which takes place in late January.

National Association of Television Program Executives
2425 Olympic Blvd., Suite 550E
Santa Monica, CA 90404
(310) 453-4440

If you're interested in promotion, check out PROMAX International. Their annual convention is in early June. PROMAX used to be the Broadcast Promotion Association (BPA), then it became the Broadcast Promotion Marketing Executives (BPME) for a few years, and now it's:

PROMAX International
6255 Sunset Blvd., Suite 624
Los Angeles, CA 90028-7426
(213) 465-3777

In the early '80s, Maria Baltazzi, fresh out of college and only two months into an entry level traffic job in a San Diego station, came up to San Francisco to the annual PROMAX (BPA) convention. She met promotion directors from TV and radio stations around the country, and befriended a dozen or so of them. After the convention, she called her new friends on a weekly basis, and within just a few weeks, she'd landed a job at San Francisco's NBC station. She had no experience in promotion but made up for it in desire and attitude. After several successful years in the San Francisco market, she was hired to produce for "Eye on L.A."

If you're interested in a career on the technical side, contact:

Society of Broadcast Engineers
7002 Graham Road, Suite 216
Box 20450
Indianapolis, IN 46220
(317) 253-1640

The place to be to learn all about cable and meet the movers and shakers is the May National Cable show put on by NCTA. They offer a publication called *Careers in Cable*. Contact them at:

National Cable Television Association Inc.
1724 Massachusetts Ave., NW
Washington, DC 20036
(202) 775-3550

Yes, there's more, and this list is not complete. You wouldn't want me to do all the work for you! Here are other organizations that may be very helpful in your mission.

AWRT has a career line for members, academic membership rates for students and teachers, an annual convention in early June, and is open to both sexes! They have some 40 chapters in markets around the country.

American Women in Radio and Television
1650 Tysons Blvd., Suite 200
McLean, VA 22102
(703) 506-3290

AAJA's annual August conventions are very popular because they include a three-day job fair attended by some 100 recruiters from around the country. AAJA also has a year-round job bank service with resumes on file, and publishes a weekly newsletter listing job postings across the country. Anyone can subscribe.

Asian American Journalists Association
1765 Sutter Street, Suite 1000
San Francisco, CA 94115
(415) 346-2051

NATAS oversees 17 regional chapters which cover 97% of the US but does not have an annual convention. The New York office, does, however, offer a number of nonpaying internships to college students interested in learning more about the industry. The Academy of Television Arts and Sciences in Hollywood, at (818) 754-2800, puts on the national prime time Emmy Awards. Local Emmy Awards are coordinated by the local NATAS chapters.

National Academy of Television Arts and Sciences
111 W. 57th St., Suite 1020
New York, NY 10019
(212) 586-8424

NABJ offers student rates, a jobline for members, internships and college scholarships. Their annual convention is in July or August.

National Association of Black Journalists
11600 Sunrise Valley Drive
Reston, VA 22091
(703) 648-1270

The granddaddy of them all, NAB's annual convention in Las Vegas draws more than 70,000 people! However, it's not conducive to networking. More likely to overwhelm. I know. I've been there. They do have student rates, however, and they do have an employment clearing house.

National Association of Broadcasters
1771 N Street, NW
Washington, DC 20036
(202) 429-5300

NAHJ's annual June convention, for which student rates are available, also includes a job fair. Their members receive a newsletter with job postings and have access to an (800) job referral line.

National Association of Hispanic Journalists
1193 National Press Building
Washington, DC 20045
(202) 662-7145

SPJ, an association serving both print and broadcast journalists, has professional and student chapters and encourages students to attend their annual convention in September or October. They also have an internship and a scholarship program.

Society of Professional Journalists
16 S. Jackson
P.O. Box 77
Greencastle, IN 46135
(317) 653-3333

WICI is a wonderful networking resource for women in all areas of communications, with local chapters and a national jobline. Student membership rates and associate rates for those in the business less than two years.

Women in Communications, Inc.
National Headquarters
2101 Wilson Blvd., Suite 417
Arlington, VA 22201
(703) 528-4000

Find out if the organizations that interest you have local chapters and about their meetings, the subjects covered, their speakers, special seminars and social events. Take advantage of all of it.

You'll be amazed at the contacts you can make and the amount you can learn by getting involved in these professional organizations, especially if you participate in your local chapters.

TEN WAYS TO FIND OUT ABOUT JOBS

- *Job lines—company or association*
- *Informational interviews*
- *Trade publications, editorial and ads*
- *Job fairs/Media days*
- *Association newsletters*
- *At association conventions or local chapter meetings*
- *By calling people who work at the station/network/cable operation*
- *Electronic bulletin boards (online conferences)*
- *Business pages of local papers*
- *Through friends. Or friends of friends.*

Make the Phone Work for You

Mike Gaynes, in his job hunting days, says it was the phone that got him in the door on a number of occasions. He explains, "The phone is your best weapon. I wasn't long on qualifications but I worked the phones like crazy. I called news directors. If you try to reach ten small or medium market news directors, you might make a friend of one out of ten. If you make five friends out of 50, you'll hear about jobs. If they don't have them, they'll tell you, and they'll tell you who does."

Knowing when to call and when not to call is extremely important. When Wendy Burch was news anchor and news director in Eureka, people would call her at 5:30 p.m. while she was anchoring and then leave a message for her to return their calls. For her, that crossed the line from aggressive to obnoxious.

Professor Hewitt says the best time to call a news director is between 11:00 a.m. and 1:00 p.m. In any event, if the station has newscasts at 5:00 or 6:00, do not ever call after 3:00 p.m. That insensitivity to a news director's job exhibits a gross lack of intelligence or consideration, or both.

You should also be aware that TV news directors are particularly swamped during the sweeps, trying to get the news on the air. (See chapter 14 for more on sweeps and ratings.) If they don't have jobs open, they're not going to want to talk to you in February, May and November.

If you do get a news director or department head on the phone, the first words out of your mouth after your name should be, "Do you have a moment?" If the answer is no, hang up and call back later.

When making phone friends, you want to monitor your personality to some extent. Lisa Heft explains one of her pet peeves, "It's hard to define what makes some people so creepy, but there is that kind of person who is cloyingly friendly, who assumes a kind of familiarity with you. There are some people who are instantly chatty who you just fall in love with, but many people can't pull it off. I'd say make sure you can, if you want to go that route. Make sure that your charm is actually working and not overpowering people."

Keep track of all your phone calls and conversations. It's all part of making a job of finding a job, according to Professor Hewitt. He recommends that you "log contacts. You want to impress people. If you remember the details of your previous conversation, they'll be impressed." Log when you called, what you discussed and advice you were given, if any. When you call back, you can say, "When I spoke with you on July 10th, you suggested I talk to so and so." You can let them know you took their advice and ask if they have any knowledge of openings. Hewitt suggests you talk to news staffers, producers and writers on an informal basis. Ask about their job market. In addition to content, date and time, note if they were cordial, the names of their kids, any information that might come in handy in future conversations.

Keep calling the job lines. If you're calling recorded messages, you can call on weekends or early morning and get the cheapest rate. But don't be cheap about the phone. It's the cost of doing business and it's considerably less than hopping on a plane.

One more word about phones. *You* need to be easily reachable. You should invest in either an answering machine or the phone company's Message Center service. You want to be able to access messages from wherever you are. Many times I was asked to find someone for a temporary assignment on very short notice. I couldn't wait until the next day for a call back, so I would keep trying people I'd recently interviewed until I was able to reach one. Several years ago, I got a call from an NBC News publicist in New York who had screwed up and asked me if I could get a still photographer to meet an NBC News crew at the Russian Consulate in 45 minutes. Correspondent Jessica Savitch was conducting an interview for an NBC White Paper special and they needed stills. Unable to reach the freelance photographers we'd used before, I called Henry Moore, a photographer who'd just shown me his portfolio the previous week. He answered the phone, grabbed his camera and film, jumped in his car and took the photos for NBC, turning over the unexposed film to them on the spot. For the next eight years, I hired Henry exclusively, because he was always just a phone call away.

ALLIES

Many an aspiring reporter, anchor, newswriter and desk assistant spends many a sleepless night plotting to get past the news director's secretary or administrative assistant. Well, guess what? The assistant's job, whether it's assistant to the news director, general manager, program director or general sales manager, is to protect the boss from the likes of you. However, that person can be of enormous help.

I asked the "protector of KRON's news director," Rosemary Wesela, about her years of screening calls in a San Francisco newsroom. Here's her advice about getting to her boss—or any boss, for that matter. She'll also tell you what not to do. Because you don't want to tick Rosemary off.

Rosemary says,

People in a market this size don't have time for informational interviews, so you have to find a way to break through. The best way to get through to a news director is to call up and ask for advice on how the news director would like this handled. And I don't mean just this news director. Any news director you're dealing with will have someone screening people, whether it's a secretary or someone on the assignment desk. The news director is doing a thousand things, so you call up and say, 'What do I need to do?' and then pay attention to what you're told, because they'll tell you. And the answer might be 'You don't get through to see the news director. You need to send a tape and a resume.' The answer might be, 'You're just out of college. He won't see you because we don't hire people right out of college,' so you might need to go to a radio station or a small market to get some experience. You may need to go somewhere you don't want to go, and that's what paying your dues is all about.

The key is to ask how you get through, ask what you need to do and then pay attention to what you're told. And most of the time, you're not going to want to hear the answer. It's not going to be satisfactory because you can't get in. You're getting bad information if you're being told to just keep calling, just keep trying, just keep bugging the news director. That isn't the way to do it. *There's a fine line between being a pest and being persistent.* If you are a pest, you could be the best thing to hit this market since the Golden Gate Bridge and no one will see you because they don't want to deal with you. You have to be confident, but you have to know when a 'no' is a 'no.'

One technique that never works is just showing up and saying 'I have an appointment.' It's as if they think the news director is brainless and not going to remember that he doesn't have an appointment. If you show up saying 'I don't have an appointment, but I would like to see the news director,' you won't get in either, nine out of ten times. In this market, you're not going to get in at all. In a small market, the guy might be free for a few minutes and would be happy to see you. So always call ahead and ask for an appointment.

One thing that I always do in screening calls for the news director is pay attention to how the caller treats me, because I know that, as a reporter, that's how they're going to treat people. They want something from me, and I pay attention to the techniques they use to get the information they want. If they are rude to me and snippy or they don't treat me well, then I know that if they're out to interview the head of a corporation for Channel 4, and he has a secretary, that's how they're going to treat that secretary.

You have to go about getting the information you want the way you would as a reporter. A sense of humor always works. Use whatever techniques you can. Do it the way you would if they sent you out on your first big story. Don't treat the little people like slobs because you'll never get through. And if that little person tells the news director that they've been treated poorly, the news director's probably going to ask, 'Why would I want to see that person? They don't know how to treat people.' So it does backfire.

There are people who send bribes. I've gotten flowers, I've gotten food, candy. It's not going to get you in any sooner than if you had just called and asked, 'How do I get in to see the news director?' I feel sorry for people who send bribes, because I feel they've gotten some pretty bad information and they don't have common sense.

Some people send things directly to the news director: plants, flowers, balloons. That isn't going to help. What's going to help is your tape and your resume. If you are looking for a job as a reporter and you don't have a tape, don't apply to a large market because nobody in a large market is going to have time to train you. By the time you get to the fourth or fifth market, you need to hit the street running. Unless there's a trainee position you've heard about, don't waste your time. Go and learn your trade.

> Your best resources are the people who work for the news director or the assistant news director, because if you enlist their help, they'll tell you what to do. They know the person, they know the person's habits, they know how you can get your foot in the door. If you enlist their help rather than try to get past them and trick them, you'll be helping yourself. A reporter named Mark Mullen called up and said, 'There's a possibility I might be moving to San Francisco, what do I do?' I told him what to do: 'FedEx your tape and resume because there's an opening. Get it here right now.' He sent it and the next day he was here for an interview. He got the job. He started out as a reporter two years ago, now he's anchoring. He asked me what he should do and he paid attention.
>
> Although I'm there to 'screen' callers, I feel very strongly that the next Diane Sawyer or Tom Brokaw may be on the other end of the line. I want you to get through as much as you do, but I must follow the guidelines set by the news director. Remember, I'm on your side. That's why I tell you what to do to get through to the boss. I know what the news director likes. Use my advice to help yourself, even though it may not be what you want to hear.
>
> If I say, 'There's nothing now, try back in six months,' try back in six months. But don't call me in two weeks and say, I called you two weeks ago and maybe something opened up.

Rosemary may sound like a tough cookie here, but she can be your biggest ally. In the 15 years I've known her, I've seen her help a lot of people. If you want to break into a TV job, treat everyone who works at a station or has connections as the expert s/he is. Make them all your allies. Befriend the friendly ones, but always respect their time. If they say you can call them back to check with them, call, but always be brief and to the point. Respect the fact that they are probably extremely busy, and don't expect them to remember you.

I always told people they could check back with me and really appreciated the ones who said, "Hi, Linda, I'm so-and-so. I'm the one who met you at the AWRT luncheon who graduated from Mills and had the pet iguana. I'm wondering if you've heard of any publicists' jobs lately. No? Well thanks, and can I check back with you in a couple of months?"

Libby Moore, Maury Povich's personal assistant, echoes Rosemary's advice, "Be good to the assistant. I'll be very helpful if you're nice but not fake, and you're honest about what you want. Don't pull attitude on the phone. How you handle yourself is very important."

10 REASONS TO PICK UP THE PHONE

- *Call job lines to check job postings*
- *Call to ask for an informational interview*
- *Call to follow up on a resume and cover letter*
- *Call to investigate more about a station, company, department*
- *Call to check addresses and spellings of names*
- *Call to check back with networking contacts to find out if they've heard of any openings*
- *Call to reconfirm your interview*
- *Call to find out about internships or minority trainee positions*
- *Call to ask a department head's assistant for advice on getting in*
- *Call to say thanks for the interview*

Informational Interviews & Road Trips

INFORMATIONAL INTERVIEWS

The idea behind informational interviews is that contacts are the key to getting a job in any industry, and you can learn a lot about the business while expanding that network of contacts. According to Right Associates, a San Francisco firm offering career transition workshops, only 20% of jobs are acquired through ads or employment agencies. Informational interviews are an important method of promoting yourself and uncovering the hidden job market. I've mentioned this hidden job market a couple of times and it's important to emphasize. You have to be a sleuth to find out about jobs in TV. It's extremely rare that a job opening in television is listed in the classifieds.

When you set up a series of informational interviews in other markets, and take your act and resume on the road, you're on a road trip, discussed in more detail later in this chapter.

Sometimes it's easier to get an interview with someone who *doesn't* have a job opening, as there's no pressure to hire you. Getting informational interviews depends on the targets of your interviews, how busy they are and how much they like to talk. Even some extremely busy people enjoy talking about themselves/their successes/their jobs and are willing to add an hour onto an already long day to help, entertain or impress you with their personal tales of the exciting world of television.

The most successful technique is to call up department heads, tell them you're very interested in the business and if they could just spare 10 or 15 minutes, you'd like to ask a few questions about how they got to where they are today and about their prognosis for the future of the industry. Tell them that you realize there probably are no openings at the moment. You're just seeking some information.

During the course of the interview, you ask them to take a look at your resume and let you know what they think. Are your skills adaptable to the business?

Before leaving, ask for recommendations of others you can call for more information. You're looking for names of other important people in the industry who can give you information, more names, and, ultimately, a job.

When you call the contacts you've been given, you're no longer cold-calling. You say, "Your friend Joe Newsdirector at WTVT suggested I give you a call. He thought perhaps you might be willing to give me just 10 or 15 minutes of your time. He said you'd be an excellent source for what's going on in TV news here in Ohio. I'm interested in learning more about blah blah."

Your 15-minute interview usually expands into a half-hour or an hour. However, be extremely sensitive to the interviewer's schedule. Don't assume you can just go on and on. After a half-hour, if you haven't been tossed out due to the interviewer's incredibly busy schedule, make an offer to leave. "I know you're a very busy person and I don't want to take up any more of your time." It's a gamble. Of course, you don't want to leave. You want to become bosom buddies and develop a rapport that will, in the long run, last a lifetime, and in the short run, turn into a job offer. At the same time, if you acknowledge how busy s/he must be, you'll be perceived as a wonderful, sensitive person. If the interviewer seems willing to continue, you can ask if s/he has time for just one or two more questions. In many cases, people love an audience, even an audience of one, and will go on and on just to hear themselves talk. (I myself have been guilty of this, which is why I sometimes sport a T-shirt that reads, "I'm talking and I can't shut up.")

If you're unable to get an appointment with the news director or head of a large department at a station, try for a middle manager in the department, perhaps the assistant news director, the producer of one of the newscasts, or absolutely anybody in the department. If you befriend a staffer or two, you'll have allies in helping you get a foot in the door.

The ultimate example of seeking allies happened one day when I was giving a tour to an inner-city class of 7th graders. Three enterprising 12-year-old African-American girls, who were sure they were the next Supremes, kept begging me to introduce them to someone for whom they could audition. Unfortunately, we had no openings at that time for Motown trios. Unwilling to give up, I caught them at the end of the tour in the lobby of the station, performing for the security guard.

So get the audition (read: informational interview) wherever you can.

In fact, here in the hectic '90s, Lisa Heft has even settled for informational interviews over the phone. Yes, you do lose the opportunity to sing for the security guard and have that face-to-face, up-close-and-personal contact, but there are certain advantages. You don't have to iron your interview skirt or shirt, and you don't have to worry about a bad hair day.

Lisa has had great luck cold-calling people in nonprofits or businesses where she might like to work. She says, "If you're brave and bold enough, you call up and say, 'I'm really interested in your field. I've done a little research.' You explain a little bit about what you've done, so you sound a little bit informed. You find the right people. You share a little about yourself. You let them share a little about themselves. And you can get that interview right there on the phone.

"It can often happen that way. They don't have a lot of time and you can often get that interview. I always thought that the person-to-person contact was all-important, but I've since found out that if you give people the option by asking, 'Is there a time you'd like to meet in person or shall we talk over the phone?', they often say, 'Well, you've got me right now. Let's go.' And they appreciate the option."

If it's handled correctly, you can learn about the company, the industry, the individual, make a good impression, and get names of others who might have jobs or leads. Many an informational interview has turned into a job. The interviewer who is particularly impressed with you may even create a job for you, or will at least keep you in mind when something opens up.

Read chapter 16 for more detail about the interviewing process.

ROAD TRIPS

If you're serious about working in TV in any department, on-air or behind the scenes, road trips are the way to find out what's out there and if it's for you.

The truth is that you need to see and be seen. A news director or program manager in another market may give you an informational interview but probably isn't going to hire you sight unseen. Unless your resume is extraordinary, you need to meet face-to-face with prospective employers to make an impression. And you also need to see the small market station and town to decide if you'll be able to adjust.

So you plan a road trip. Before leaving home, do your homework. Get out a map. Decide which direction you're headed and what TV markets you'd like to hit. Go to the library and copy the pages from the *Broadcasting & Cable Yearbook* that list the stations in those markets. Determine the stations that are of interest and look up the names of the people you'd like to meet, the department heads of the areas that interest you. Call the stations to verify that the person listed is *still* in that position, that you've got the name spelled correctly and the correct title. Just because you found a name in the current *Yearbook* doesn't mean it is still accurate now, or even when the book came off the press.

Send your resume and a cover letter to the department heads you've targeted and indicate that you'll be in their town on such and such a date and would very much appreciate meeting with them briefly. Follow up with a phone call to confirm.

Many small station news directors and department heads will meet with you if you're making the effort to come all the way out to their neck of the woods. TV people respect job seekers who take initiative.

Plan to arrive in town the evening before your meeting so you can watch their news and other station programming.

Professor Hewitt advises, "Don't be stupid. Know a little about the community. Know the station. Jot down the names of their anchors. Xerox the pages from *Broadcast & Cable Yearbook* on the markets you're going to visit and familiarize yourself with all the stations, including the competition. Use those research skills."

When you return home, write thank you notes to all the people you met, and follow up regularly with phone calls to them. You not only want to know if an opportunity has opened up in their department, but also if they know of any other leads. Handwritten thank you notes have a nice personal touch and give the employer a chance to see if you can spell without a spell-checker. If you have a computer, a typed letter is also fine and has a professional feel to it. Frankly, interviewers are impressed by any thank you note. Thank you notes not only give you a chance to sincerely express your gratitude, but also a chance to follow up, make a good impression by writing an intelligent letter and reiterate your interest in the company.

If you go into your job search with realistic expectations of the commitment required, you won't be easily discouraged. In fact, you'll maintain that all-important positive, confident attitude and you'll make something happen for yourself.

TEN WAYS TO MAKE A GOOD FIRST IMPRESSION AT AN INTERVIEW

- *Arrive 10 minutes early*
- *Be enthusiastic*
- *Be relaxed, self-assured, confident*
- *Look neat and professional*
- *Be courteous to the receptionist/secretary*
- *Have your resumes with you, without typos and professionally printed*
- *Know the background of the company*
- *Ask intelligent questions*
- *Don't make negative remarks about past jobs or bosses*
- *Finally, but most importantly, ask not what the company can do for you; ask what you can do for the company*

14

Talk the Talk

Do you know a sound bite from a spot, a stand-up from a slug?

Familiarizing yourself with the language spoken in the halls of television will help you make the most of informational and job interviews. It could mean the difference between your gaining an ally and perhaps a job, or being back on the street with a red face. You've got to talk the talk before you walk into the interview.

People making the hiring decisions expect applicants to be familiar with the workings of the business. You can't compete if you don't know the language. In my interview for the first TV job I eventually landed, I knew very few of these words. I survived the interview only because I pretended to know what my future boss was talking about. I nodded at the right times and laughed at the right times. It's risky business. I don't recommend it.

Following are just a few key words bandied about in everyday TV conversation.

Market. TV markets are geographically defined viewing areas. According to the 1994 *Broadcasting & Cable Yearbook*, there are 209 markets in the U.S. They are ranked in size based on the number of television homes in the area. The nation's number one market is New York, made up of more than seven million households. The 209th market is Alpena, Michigan, boasting some 15,000 households. Market size doesn't correspond with city size since markets reach beyond city limits and sometimes over state lines, often bunching several cities and towns together if they share geographic proximity. For current market rankings, see the Appendix.

HUT levels. HUT means "homes using television." It's the percentage of homes with their TV sets on. HUT levels are highest in the evening, since many people are in school or at work during the day. HUT levels are usually higher in the winter when cold weather keeps people from enjoying outdoor activities. Also, not surprisingly, HUT levels soar during coverage of special events such as "The Academy Awards" or Super Bowl or when major news is breaking.

Ratings. A rating is the percentage of total TV homes in a given area watching a particular show. So if you say that WXYZ pulled, drew or scored a 10 rating for their 6:00 p.m. news on a given night, that translates into 10% of the homes in the market tuned in to that particular newscast. Ratings are a life-and-death matter to programs and newscasts and the people responsible for them, as the revenue of the station or network is directly impacted by them. The commercial rates paid by advertisers are set, not only by supply and demand, but by the number of eyeballs the station estimates it can deliver based on the past performance of that show or a similar show. The TV station actually *sells* its audience (potential consumers) to advertisers. The A.C. Nielsen Company is an independent research firm that delivers the ratings on a daily and monthly basis to its subscribers, which consist of TV stations and ad agencies. The overnight ratings are measured in two ways—by electronic boxes attached to the Nielsen families'

TV sets and, during sweeps, by diaries that are filled out by a larger sample.

Share. A program's or newscast's share is the percentage of the HUT level (homes with TVs on) that are tuned in to that show. To determine the share, divide the rating by the HUT level.

Gross rating points (GRPs). When a spot (commercial, promo, PSA) airs multiple times, the GRPs are determined by adding the rating of each airing. Essentially, GRPs indicate the total exposure the spot receives.

Demographics. Ratings can be broken down into demographics—characteristics of the viewing audience by sex, age, income and ethnicity. This information helps sales researchers and AEs work with media buyers and clients to determine the best times for an advertiser to reach the folks most likely to buy his or her product or services. If a TV account executive can convince a media buyer that a particular show delivers the ideal demographic for a particular product, s/he will be able to negotiate a higher rate for the commercial time than the straight rating would warrant.

Book/sweeps. The TV ratings which most affect advertising rates are measured during 4-week "sweeps" or ratings periods. The three most important sweeps are in February, May and November each year, always beginning on a Thursday near the beginning of the month, or the end of the previous month, and ending four weeks later on a Wednesday. Shortly after the end of a rating period, the "book" comes out. A.C. Nielsen delivers the data for the sweeps in a ratings book and the TV sales research team goes to work to evaluate the results in the best possible light.

Spots. Another word for commercials, promos and PSAs. Most spots are 30 seconds in length, although some run 10, 15, 20 and 60 seconds. Add to the mix your newscasts and updates, network, syndicated and local programming, and station IDs, and you have the elements making up the daily program log.

Talent and personalities. These two terms refer to a station's on-air employees, such as reporters, anchors, sportscasters, weather anchors and show hosts.

Focus groups. These are groups of individuals surveyed in a controlled research environment as to their recognition of and reactions to on-air personalities, programs, slogans, advertisements, etc.

Q rating. This is the recognizability factor of a station's talent. If someone has a high Q (not to be confused with IQ), it means that s/he received positive responses from focus groups.

Rundown. A rundown is the "bible" put together by the newscast producer which includes all the details the director needs in order to orchestrate the live newscast from the control room. It lists the sequence of the stories in a particular newscast, how much time is allotted to each, and which stories include video, graphics, supers, voiceovers and sound on tape.

Backtime. A producer of a live newscast or program times the program elements, counting back from the end of the show, with the help of a computer, to assure that the program ends on time. If a segment or interview runs long, the producer adjusts by shortening or dropping upcoming segments to end on schedule.

Slug. This refers to the abbreviated name given a story in the rundown to differentiate it from the rest of the stories on the show. Examples might be "Main St. Fire," "Tax Hike" or "Flood Watch."

Supers. Supers are words that are superimposed over the image. They may identify the anchor, reporter, or person speaking. They may be locators, indicating the city or location of the event, City Hall, The Astrodome, White House, or sports scores or weather information such as temperatures. They also identify time frames, such as Live or File footage or Yesterday. Today, stations and networks often "super" their logo in the corner of the screen over news and programming so that the viewer will always be reminded who they're watching.

Sound bite. Often called simply "bite" for short, a sound bite is a very short clip of someone being interviewed that encapsulates the essence of a story.

VO. This stands for voiceover and refers to the copy read while video is shown. The speaker is not seen on camera.

Sound on tape (SOT). Sound on tape refers to dialogue recorded at the same time as the video. So on a rundown, if a news story indicates SOT, the anchors know they must clam up.

Stand-up. A stand-up is a news story in which the reporter is seen on camera, live or taped at the scene.

Package. This is a news story or segment that is recorded and edited for playback in a newscast or program.

Spot news. Spot news is breaking news, something unscheduled that happens and is being covered today.

Color bars and tone. This image of color stripes accompanied by a tone at the beginning of video allows technicians to adjust the color and sound levels for air.

Academy leader. This second-by-second countdown edited onto the front of video indicates how many seconds before the beginning of the story or piece.

Slate. This information on full-screen at the beginning of video identifies it but is not aired.

10 GIMMICKS THAT DON'T WORK (OR BACKFIRE!)

- *Claiming to have an interview when you don't*
- *Dropping in without an appointment*
- *Pretending to be someone you're not*
- *Flowers*
- *Candy*
- *Booze*
- *Threats*
- *Lies*
- *Anything that's not clever*
- *Anything inappropriate to the company or department*

Gimmicks

A few job seekers resort to gimmicks to get the attention of a potential employer. The most common result is that you win the battle, but you lose the war. In other words, you do usually get the attention of the recipient of your ploy, but the overall effect is often negative. Whether or not gimmicks work depends on whether the recipient thinks they're clever or appropriate. Some departments, by their nature, are more gimmick-oriented. Promotion and sales come to mind. News, on the other hand, takes itself a little more seriously and there's something about journalistic ethics that puts a damper on jokes and tricks.

Here are some examples of gimmicks people have used to get an interview or call-back.

Someone who shall remain nameless gave me this trick for getting through to the department head. When the secretary asks, "May I tell him what it's regarding?" in her attempt to screen you out, you reply, somewhat frantic, "Well, I just wanted to let him know that his wife found out about us." My friend claims that should get you right in. I don't recommend it. In my opinion, few department heads would find this amusing.

Joe Fragola, when hiring for a couple of dozen jobs for BayTV, was called down to the front desk. A box had been delivered for him. Inside was an apple pie with one piece missing and a little flag saying, "KRON is missing a piece of the pie." The next day, he received a smaller box with his name in the same handwriting. Inside was the missing piece of pie with a woman's name and phone number on little flags.

It so happens that Joe likes people who take chances, so he responded. He said, "She's a publicist who started out in broadcasting. She's produced video news releases so she has a sense of what we do. She's willing to give up a good paying job to come work for us as a production assistant."

Joe not only interviewed the applicant, but subsequently hired her as a producer/writer/update anchor for BayTV.

About the same time, Joe received a box from another aspiring journalist. This one contained a bag of water. Floating inside the bag was a Calistoga water bottle. The bag was tied with a ribbon and bow, and the tag read, "I want to make a splash." Rolled up inside the bottle were a couple pieces of paper: the applicant's resume with a great cover letter explaining why she wants to make a splash, and, borrowing a page from David Letterman, the top 10 reasons why she won't aggravate Joe. She included three postcards addressed to her, one with a guy dressed in a monkey suit saying, "I don't want to monkey around, call me for an interview." All three postcards were positive responses and all were signed by "Joe."

This one didn't land a job.

A couple of weeks later, Joe received a box wrapped in money wrapping paper saying, "Holy Shit, do I want to work for you!? I'll work like a dog. I'll do anything." It was signed Bobby Brownnose and inside was a plastic bag full of cat poo and dog biscuits. It happened to be April 1st. The culprit hasn't been found, but Joe suspects one of his cohorts in the newsroom.

One job seeker had a pizza delivered to a news director one noon. The center of the pizza had been replaced by his resume and a note, "Lunch is on me. I just want four minutes of your time."

Another job seeker sent one of her shoes to a prospective employer, with a note that she just wanted to get her foot in the door.

One applicant for a publicity job sent me a basket of customized fortune cookies, each one extolling one of her qualifications. I thought it was clever and granted her the interview, although I didn't hire her. Her gimmick was appropriate for a promotion job, as getting attention is what promotion is all about.

Sandi Ball, a TV sales account executive, sent a series of balls to a prospective employer, some of them sandy. They were labeled with appropriate puns, such as, "Take the ball and run with it," "The ball is in your court," etc. She *did* get the job. She exhibited that she had a lot on the ball and would have the kind of personality that would help her get the attention of the ad agencies' fickle media buyers.

A fellow with a publicity background applied for the coveted position of researcher on "Bay Area Backroads," the top-rated local non-news program in the Bay Area for many years. He made a black-and-white photostat of a map of the Bay Area, glued it into a box and pinned on about 200 little blue flags, one at each little town and crossroad around the Bay Area. The little flags all said, "Been there," "Been there," "Been there, too." Mounted in one corner of the 18x 24" box were his resume and cover letter. The entire project was fastidiously and beautifully put together; a real work of art.

Alas, the bottom line is that he didn't land the job. The person who had been temping in the position ended up getting it. The temp had had a chance to prove that he could do the job, already knew all about the station and its operation, and was promoted from within.

I, however, was very impressed with the applicant's unique resume box, and, noting that his publicity background actually made him a candidate to work in my department, I took him to lunch to let him know that his efforts were appreciated, see if I could give him some job leads and let him know I'd keep him in mind if something opened up in my department. I also used his map/application for several years for show-and-tell at career panels. I even used it as an example of a clever resume on the local CBS affiliate talk show "People Are Talking." I was invited to appear to discuss, of all things, gimmicks people use to get their foot in the door.

Gimmicks do make the recipient pause and take note. If the idea is original and perceived as clever *and* you have the qualifications for the job, you'll probably get an interview. If the idea is really exceptional, and you don't have the qualifications, you may get an interview as a reward, but you won't get the job. However, you might be recommended for other jobs for which you are qualified and they could turn into a productive informational interviews.

But before you resort to a gimmick, seriously consider if it's appropriate and clever, or if it could backfire.

Flat-out bribes—candy, flowers or a bottle of booze—are highly frowned upon. Most employers resent the pressure and feel these kinds of gifts are inappropriate and, frankly, tacky.

On the other hand, tasteful thank you notes after an interview are not considered bribes or inappropriate.

Getting Past Security: You're In!

16

The Interviewing Process

PRE-INTERVIEW PREP

Getting the interview, informational or job-specific, may be tough, but it's just the beginning. Organization is essential. Prepare yourself for the long haul. You may think you have a terrific memory and that you'll never forget the details of your conversations, but trust me, you will.

Thus, in addition to a phone log system to keep track of who you talk to on the phone, who they refer you to, what you discuss, and when you called back to check in, I highly recommend an interview checklist. You can develop a form to your own taste, print out hard copies and fill one in for each interview, not *during* the interview, but before and after. During the interview, it's appropriate to have a pen and pad of paper to jot down names, numbers and brief notes. If you're set up on computer, all the better. Keep your records on disk, but print out the data for easy reference when you're out and about.

Here are my suggestions, a starting point for your list.

- *Date of interview*
- *Time of interview*
- *Interview is with (check correct spelling)*
- *Title*
- *Call letters of station*
- *Channel number*
- *Affiliation*
- *Address of station w/ zip (for follow-up and thank you note)*
- *Directions to get there*
- *Assistant's name*
- *General manager's name*
- *Newscast times*
- *Other local programming*
- *Other people I know or with whom I've spoken at the station*
- *Other background on the station*
- *Questions I have for the interview*
- *Important points I want to make*

Immediately after the interview, write down your impressions for your records and follow-up on these subjects.

- *General advice given*
- *His/her greatest needs/challenges*
- *Overall impressions of station*
- *Degree to which this person is on my team*
- *Referrals from interviewer*
- *Other contacts I made while at the station, with notes about them*
- *Follow-up plan*
- *Date thank you note sent*
- *Date to call this contact to follow up*

THE INTERVIEW ITSELF

The moment of truth has arrived. Have you? My point here is that if you're late to your interview—for any reason, short of death—you've just blown it. Perhaps nothing is said, but, believe me, it's been noted. TV runs on drop-dead deadlines, as in, "The show will go on at 6:00:00, *not* give or take a few seconds." People in the business are only interested in hiring people who do whatever it takes to get to a story, the control room, the fire, the set, on time. Allow plenty of extra time to find your way, get stuck in traffic, get a speeding ticket, find parking.

Job hunting becomes infinitely easier (as does so much in life) if you simply adhere to the seven P's of life. To wit: Proper prior planning prevents piss-poor performance.

Have you planned properly? Are you prepared? Have you filled out your interview checklist? Have you looked it over? Do you have your resume? Do you know where you're going? Have you taken four deep breaths in front of the mirror, affirming "I'm the greatest" on every exhale? Have you checked your teeth for remnants of this morning's cereal?

What to wear? Bill Groody advises dressing like the people where you are applying. Not over or under. You might make an exception to that rule if everyone's in T-shirts and jeans. I wouldn't be quite that casual.

Have you done your homework? You've been reading *Electronic Media* and *Broadcasting & Cable* magazine. You know the trends, the players, and you can talk to people on their own level.

As a news director, Doug McKnight has hired a lot of people over the years. He can't say it often enough: "If you get to the interview stage, find out everything you can. Don't come in and ask what network we're affiliated with or what time our newscast is on. My attitude is that research is one of the skills I'm looking for and if you didn't take the time to research my station and know something about me and the place I represent, then you really don't have that much interest in working here and you're probably not going to do a good job as a reporter researching a story. So take that time. It's absolutely essential."

Lisa Heft's two cents' worth: "I like interviews where interviewees come with some interesting things to say. They did a little background work. They offered information about themselves that was germane to the conversation, not just pushing in more extra info. They are presentable, well groomed, they greet other people. I really notice if they greet the front-desk person. Whoever they see, they've got to respect. Because then I can see what they would be like out in the community representing our station."

Perhaps the most important advice comes from Lisa Simpson with NBC News in New York: "Present yourself as a solution to a problem. Understand the needs of the organization, then tailor yourself to meet a need. Demonstrate that you really 'get it.' Be up on current events, demonstrate you know what's going on. It's more than a job, it's a vocation. It's very competitive. You have to be so conscientious. There is so little original thinking. In an interview, I'd love for someone to lay out their philosophy for me in terms of where TV news is going, whether or not I agree with it."

At the risk of stating the obvious, the bottom line is the bottom line. TV stations are in business to make money. Period. This is an important point to keep in mind when you're interviewing. Keep asking yourself, "How can I help this station make more money?" I know it sounds crass, but I can assure you that that's what they're asking themselves as they look over your resume.

John Hewitt says you should be ready for two questions in news interviews. Where do you want to be in five years? What would you criticize about our news?

This second question calls for finesse. One of the biggest mistakes you can make is coming into an interview telling the powers that be how badly they're doing their programs, newscast or promotion. You want to present yourself as a "solution" without telling them they have a problem. They may, in fact, have a problem, but they're not going to want to hear it from you. One of my personal pet peeves used to be having people tell me I simply couldn't live without them. I obviously had, for some time. There are ways to show what you can bring to a department without implying interviewers and their stations will fail should they pass you by.

Your interview should be a give-and-take exchange, with both you and the interviewer asking and answering questions. If the interview is for a specific position, your questions will include more specifics: What are the position's greatest challenges? What are the most important attributes for success in this area? What is the turnover rate for this position and why? What are the opportunities for advancement within the company, and what's the natural progression from this position? What computer systems are used in this department?

For both job-specific and informational interviews, you may want to ask: What future trends do you see in this business? What is the station doing to adapt to these trends? To which professional organizations do you belong? Which do you recommend for the people in your department?

In the informational interview, you ask more about the individual: How did you get into this line of work? What are the most rewarding and frustrating aspects of your job? What would you recommend for someone starting out in this field today? How much creativity and autonomy do you have in your department? What is your greatest need? What do you look for in a prospective employee? What are the entry level jobs in your department? Would you take a look at my resume here and give me any feedback on it? How might my experience apply to this field? What could I do to improve my resume? Who else would you recommend I call for additional information?

If you're applying for a specific job, you'll no doubt want to deal with the issue of money. If the job is entry level and you're competing with hundreds of others, you won't have much bargaining power. Perhaps the salary was posted along with the job. If not, you may want to inquire. However, that winning can-do attitude so appealing to employers includes a desire to work long and hard with no concern whatsoever about pay. Being overly interested in the salary, raises, overtime, benefits and perks is a turn-off. I've known more than a couple TV *bosses* who've had the gall to say, "You should pay *me* to work here," and some of them mean it.

Ideally, you want interviewers to decide they have to have you for the job. At that point, you may have some negotiating power, but tread lightly if you are serious about getting your foot in the door.

POST-INTERVIEW

It bears repeating: Get thee to a computer and get those notes down, with your follow-up plan. Who else to call for appointments? When to call the interviewer back?

What advice did you get in your interview? What can you do to follow up to show that you were paying attention and to rise above the competition?

One way to show your interest and sincerity is to sign up for a class to improve in some of the areas your interviewer suggested, a technique I used successfully twice, as I mentioned in chapter 5. You not only improve your qualifications for a job, but you show the prospective employer that you are willing to do whatever it takes to get the job and then get the job done.

Thank you notes for interviews—informational and job—should always be sent immediately. Promptness reinforces that you're someone who's responsible and thorough. I haven't known anyone who didn't appreciate being appreciated. Your note should be something along the lines of, "Thanks so much for meeting with me yesterday to talk about the publicist position in WJOB's promotion department." You might refer to something you found particularly informative, interesting or helpful, and reiterate your interest in the station or job. You need to walk that fine line between sucking up and grossing out. A sign posted on my office wall for many years says it best: "The key to success is sincerity. Once you can fake that, you've got it made."

Selling Yourself

In March 1992, Bill Groody spoke on a panel at the Northern California RTNDA convention about getting "A Foot in the Door." His career spans TV and radio, including WJZ in Baltimore and NBC Radio in Washington, DC. Groody's insights into selling yourself warrant repeating here, and I do so, with his permission:

> In this tight job market, one of the most important things is, how do you sell yourself? I did it successfully a couple of times, but I also blew it several times. I wanted to work for KCBS, but I could never crack that nut. I could never get in that door. Because I never really had the proper relationships and I was never really aware of the proper approach. When I bought my first radio station, I became a lot more involved in sales, and I realize now that selling yourself and selling radio time are really not that different. There are a lot of things we can learn from our brothers and sisters in the sales department about human relations, and human relations are very important. This is a people-intensive business. News is very people-intensive.
>
> You have to understand how the mind works when a person is in the process of *buying* your services—in this case, when

the news director is buying something. There are four steps in the sales process: Awareness, Interest, Desire and Action. You should not expect to get someone to hire you right off the bat. News directors or station owners want to feel comfortable with you. They want to understand that you're somebody they can trust, somebody who will work well for them.

Awareness—that's where your resume and cover letters come in. They are the first things that let the prospective employer know who you are. Nonjob-seeking contacts are important. Do you know anyone there? Do you have anyone who can get you in the door for an interview? Who knows your temperament? Do you get angry? Are you easy to work with? Are you competent? Do you write? Are you aggressive?

When you send out a resume, follow it up. At the bottom of your cover letter, always say you'll give them a call, or something like that. Then, call them with a goal in mind. That goal should be getting an interview.

The interview is the interest phase. You've obtained their attention. They know who you are. Now you want to build interest. Get them interested in you. In the interview, my advice is to ask as many questions about the station as they ask about you. It's a two-way situation. Yes, they're learning about you. They're sizing you up, but you want to learn about them. Because you want to be able to cast your abilities and skills in a way they will find most helpful. So you want to ask them a lot of questions.

As a matter of an approach, join the team of the company where you're thinking of being hired before you're ever hired. Look at the whole situation from their point of view. Ask, 'How can I help this company? What am I going to do to help this guy who's hiring me look a little better?'

The next phase is desire. You want to get them to want to hire you. You want to be eager, but not overly eager. You have to show them that you want to work there. Then, if in the process of looking for a particular job, another employer offers you a job, say, 'I really want to work for you, but somebody else has offered me a job. I don't want to take it, but what can you do?' Try to build desire by making yourself more attractive. If the employer you really want to work for turns you down, don't stop there. You might go ahead and take the other job, but continue to pursue the relationship with the place you really want to be. If you've been successful in your interviews, you know something about the person you're dealing with. Maybe he has a hobby and you see a newspaper clipping about it. You send it off with a little note, 'I was thinking about you.' It helps establish that commonality, that relationship that's going to make it more comfortable for him to hire you.

Action. When the first three steps are completed and there is an opening, and you have the skills, presumably you'll get hired. Sales is a numbers game. Increase the amount of exposure you get and you increase your chances.

Enthusiasm, knowledge and skill equal jobs. To succeed, you should be nice, have good manners, don't interrupt, be on time. If the news director is interrupted by business during your interview, let him do his business. Don't try to keep talking. Say 'thank you.' Don't get angry. Don't use negatives. Don't trash other stations. Don't trash other people. Don't share gossip, even if a news director or some other interviewer wants you to. Show confidence. Make sure you're talking to the man or woman who can hire you. And remember, the news directors are most interested in what you can do for them.

10 BENEFITS OF VOLUNTEERING (AND INTERNSHIPS)

- *You make contacts in the business*
- *You acquire skills and knowledge of the business*
- *You make a good impression on people who can help your career along*
- *You're in the right place at the right time (hear of opportunities)*
- *You gain a valuable addition to your resume*
- *You get a chance to prove yourself*
- *It's much easier than getting a paid position*
- *You don't have to worry about paying taxes*
- *You can't get fired*
- *You earn good karma*

Volunteering

If you're out of college and ineligible for an internship, you might consider volunteering to learn more about the business and make some contacts. Volunteer opportunities abound at local cable operators and at PBS stations. However, local television stations and the networks rarely even let you give it away.

Paul Dunn at WCNY, the public television and radio stations in Syracuse, New York, says, "We use volunteers in all sorts of capacities: stuffing envelopes, answering phones for pledge drives. We even have a couple of volunteers who host shows. One, a college president, hosts 'In the Loop,' an educational show. Another, an investment banker, hosts 'Financial Fitness,' a prime-time TV program which addresses viewers' financial questions. Our radio division provides a reading service for the blind, which involves more than 100 volunteers reading from local newspapers and magazines." He goes on to say you might start out compiling mailing lists for the PR department and end up with a full-time job.

Ron Bozek did just that. He volunteered at KQED, San Francisco's PBS station, helping the station publicist. After six months of volunteering, he was offered a job in the audience services department, answering calls from viewers. KQED, in fact, has a distinct volunteer services department.

All PBS stations rely heavily on volunteers during their Pledge Drives, not only behind the scenes but also on the air.

Local cable operators that have local origination (LO) departments (read all about them in chapter 23) use volunteers to help with production. In some cases, volunteers are also used as on-air talent. LO departments can offer great hands-on experience for someone who wants to learn the ropes.

Local television affiliates often confine their volunteers to internship programs for credit, citing union regulations or insurance problems as their reason. Sometimes there are legitimate volunteer opportunities, however, at a local station. Diana Mordock, after graduating with her communication arts degree from San Francisco State, volunteered to work on "Airlift Africa," a famine relief project sponsored by San Francisco's NBC affiliate. The contacts she made at the station helped her secure a paying news assistant position a couple of months later.

Occasionally volunteers are recruited to answer consumer complaint lines, and sometimes to help with special promotional campaigns. At KRON, I recruited volunteers to help when syndicated shows came to town. I needed several volunteers to work the "Live! with Regis and Kathie Lee" show when they taped a couple of shows from Pier 39, and another six or so to work with Geraldo Rivera and his staff when he taped four shows at a couple of South Bay shopping centers. These were short gigs, but when the gig was up, the clever volunteers had befriended some of the crew and added them to their network of allies in their job-seeking efforts.

Finding out about these opportunities means making contacts at your local stations and keeping in touch with them on a regular basis, because the person looking for temporary or voluntary help will call the people who come to mind first.

Sometimes volunteering in the not-for-profit world puts you in close contact with TV people and allows you to prove your worth and develop relationships. TV stations often work with nonprofits to put on major community fundraisers. The teams consist of people at the station working in concert with people from other cosponsoring media and the benefiting nonprofits to produce and promote the event together. In addition to the contacts you meet and the chance to show them how indispensable you are, you can also learn a lot about how the TV station operates. In my years in television, I met many outstanding volunteers who worked on joint projects such as walks, runs, food drives, historical events, festivals, a series of concerts to raise money for AIDS research, even a cat and dog fair.

Many stations that don't accept volunteers do allow station employees to volunteer their own time in another department. After landing a weekday job as KRON's receptionist, Aimee Rosewall worked weekends for free on the news assignment desk. Even though she was already an employee at the station, she says it wasn't easy. She "had to argue for it." She worked every day, either as a receptionist or desk assistant, for three months that first summer. A few months later, she was caught in a station layoff, but only a month after that, she got a call from the associate news director who offered her temporary work as a news assistant. While she was training for the temp job, a weekend desk assistant quit and she was offered that job. We'll pick up Aimee's saga in the next chapter. But the point here is that she found a way to volunteer and gain the experience she needed to land a paying job just a few months later. And she did it while working full-time. It's this kind of commitment that makes the difference.

TEN REASONS COMPANIES HIRE TEMPS

- *Someone leaves before a replacement is hired*
- *An employee takes a leave (maternity, travel, family business)*
- *An employee calls in sick*
- *An employee goes on vacation*
- *To fill regular positions without paying benefits*
- *For a special event—parade, dignitary's visit*
- *For a special production—telethon, sports coverage*
- *For breaking news—disaster coverage*
- *For a remote taping or broadcast to avoid paying travel expenses—Monday Night Football, "Today" or talk show remotes*
- *To try someone out before offering a permanent position*

19

Temp Work

TV stations hire a variety of temporary employees. Some of these are drawn from the ranks of experienced former employees who may have been laid off in earlier cutbacks. Others are freelancers with special skills who find casual work wherever they can. However, there are occasional temporary jobs for those with no previous experience, usually as a secretary or receptionist. Temp job schedules vary like permanent schedules in TV. Some are days, some are nights, some are weekends, some full-time, some part-time.

One factor in kick-starting a career in TV is being in the right place at the right time. Temps are in the right place, and those with TV aspirations are hoping that the right time will coincide with their tenure, however brief.

If you're looking for a full-time job and are available for temp work while you're searching for something permanent, I recommend you call the local TV stations' personnel or human resources departments and ask about working for them as a temp. If they use a temp agency, register with that agency and then use all your charm with the agency to get the plum assignments at the station.

I tried cold-calling a couple of stations to ask what temp agency they use and in one case was told they simply hired former employees. After explaining that I was actually researching this book, the human resources director explained that she, in truth, does use an agency but was reluctant to offer that information. The reason is that she gets calls regularly from reps of other temp agencies who are trying to sell her their services and want to argue about the relative merits of the various agencies in town. So if you do call, be up front about why you're calling and you'll get a better reception.

If you make yourself available directly to a station, bypassing a temp agency, you can save the station some money and perhaps make more than you would if you were working for the agency. So the station has an incentive to hire temps directly, if they have people they can count on who are easily reached and dependable. Another reason you may want to try to temp directly for the station is that some temp agencies make you sign an agreement to prevent you from taking a job at the station (or wherever you're temping) for a certain length of time after the temp job ends.

Very often, an overloaded department will get the go-ahead to hire a temp for a couple of weeks or on a month-by-month basis. Sometimes these positions are essential and the temp job continues month after month, occasionally acquiring permanent status. Even if the job is limited in scope, it affords an opportunity of learning about the business, making contacts, and proving one's value to the department while making some money.

Temps are often needed to fill in for people going on maternity or other leaves of absence, or for special projects. I've hired temps to coordinate health fairs and to work weekend events when I couldn't find enough volunteers.

Once I prescreened temps to work for two to three weeks on "Today," which was broadcasting from the Bay Area. Although the show itself was only originating in the Bay Area five days, a tremendous amount of set-up was required, renting equipment, building stages, acquiring permits. The week of the event itself, a crew of 15 or so local temporary production assistants were needed to work with the audience and guests, and to assist the producers and the talent. The "Today" coordinator on the San Francisco shoot entrusted me with finding her the best candidates. I drew from former interns, people who had temped for me before, and bright, enthusiastic job seekers I'd met during the previous year.

They were interviewed for the positions and the ones hired worked at neck-breaking speeds around the clock but loved every minute. They all ended up with NBC's "Today" on their resumes.

Network sports departments routinely hire local help for major sporting events in a city—such as a Super Bowl or an All-Star game or even regular season games they're covering. If they can hire locals, they don't have to pay travel, meal and hotel expenses.

News departments hire additional temp help for coverage of special events such as the Pope's visit to town, election coverage or a Bicentennial; and programming departments hire temps on a project-by-project basis.

All are great opportunities to expand one's circle of contacts.

Aimee Rosewall, who, as the curtain fell on our last chapter, had just turned a volunteer effort into a weekend news desk job, found that KRON wasn't keeping her busy enough to pay her bills. She managed to negotiate a temp contract through the station's human resources department and worked in just about every department, filling in wherever her talents were needed most. The station paid her a salary and then assigned her to various departments on a week-to-week basis. Eventually, a full-time news planning job opened up on the assignment desk (reading the mail and getting background on scheduled news events). Aimee gave up the temp work to join the desk, but the temp job served its purpose. It kept her name and face in front of everyone until something permanent came along.

TEMPS IN ENGINEERING

Temps are a regular phenomenon in TV engineering departments. At WLS, the ABC affiliate in Chicago, some 95 out of a total staff of 270 work in engineering. There's less turnover in engineering than in other departments, and less moving around, no doubt in part because large market stations are union shops.

Jim Owens, director of engineering for WLS, says, "A lot of people work that way all the time. Temp jobs are about the only thing we have available." Owens receives hundreds of unsolicited resumes each year, and is sorry that he doesn't have any permanent jobs to offer. One reason is more automation and efficiency. Many large stations now use robotic cameras in their studios, eliminating the need for camerapersons. As engineers retire, they usually aren't replaced. The staff is reduced by attrition and any additional help needed is hired on a temp basis. Some technicians have temped for years, making it tough even to get on the temp roster in engineering.

Caught on Camera

So You Want to be an Anchor?

Most people who aren't in the business think of on-air jobs as the most glamorous in the business. Yes, a lot of reporters, anchors and show hosts love their work, when they have it, and the excitement, celebrity and paycheck that go with it. For some newscasters, it's the desire to let the community know what's going on, filling the noble role of guardian of free speech and democracy, watchdog of the world. For others, no doubt, it's the proximity to power and fame. But many in the biz want you to know that it has its downside and it's not for everyone.

And if you think breaking into TV is difficult, try landing an on-air reporting/anchoring/host job. Certainly, you don't start out on camera. You work your way up, after learning the business. There are about five times as many jobs in newswriting and producing as on-air. And be forewarned, if you want to get into management, you'll get there faster if you're *not* on-air.

Here's Rosemary Wesela's take on on-air jobs: "A normal progression in news is internship, production assistant, writer, then producer, and if you're smart, that's where you'll stay. You have regular hours for the most part if you're a producer. You can have a life. You don't have to keep clothes under your desk because you might be called out in the middle of the night and be gone for three days. You don't have to worry about having a bad hair day.

"The pressure of this business is that no one is fire-proof. You're not going to get a job as a producer, as a writer or as an on-air person and have even a small amount of job security unless you own the station. The news director is not going to blame himself if the six o'clock news isn't working. He will look to the anchors, to the reporters and to the producers. You can be the best there is in any one of those positions, if you're on the air you can have a following of millions of people, everyone loves you, everyone thinks you're the hardest worker there is, and then a new news director comes in. And news directors' average life span is 21/2 years. He has his favorite producer of the 11 o'clock news where he came from, and you'll no longer be there. You'll be demoted, you'll be moved, or you won't have your contract renewed.

"So if you want to get into this business, to be on television, to be a celebrity, you're going to have a tough time, because you're going to lose jobs for no reason at all. And if you want to get into this business because you want to tell people what happened, then you won't mind all the other baloney because you will still be feeding the information you want to get out. You're not there to be a celebrity. You're there to dig deep. What's behind that story?"

When people tell Mike Gaynes they want to be on-air, he advises, "Don't. It's high stress, low pay, incredibly political. There's no security, no parachute for your ego." And if that doesn't discourage them, he admits, "You know they've got what it takes. If you don't love it, you'll be miserable in it. You've got to love it or you won't be a happy person. You have to feel passionate about it."

THE PLAN

Still passionate? Okay, here's the plan. Go to a college with its own TV station and get as much experience as you can there. Take a page from Wendy Burch's book and get an internship, attend RTNDA conventions and take road trips to small markets. Do whatever it takes to become the best newswriter you can. Learn how to edit, shoot and produce. Put together the best resume tape you can. (See chapter 22.)

Try to be realistic about your talent and looks. Do you have that *je ne sais quoi*? That on-air charisma? Can you hold the audience? Are you attractive enough? Bright enough? Are you well informed about current events? Are you a news junkie?

It's not going to happen overnight. But if you've got what it takes and are willing to go where the jobs are, it can happen fairly fast.

RIGHT PLACE, RIGHT TIME

Not to negate everything I've just written, but occasionally someone just falls into an on-air job. Evan White, senior anchor for BayTV in San Francisco, started by accident. He said, "In 1957, I was hitchhiking across the country and I ran out of money. I stopped at my brother's in North Carolina to eat, and his wife had done some work at a local radio station. She said, 'You ought to go down, they need a disc jockey.' I didn't know anything about radio, but in North Carolina, the one criterion they had for the job is that you didn't have a southern or an eastern accent. They wanted no accent. I was from California. I had no accent. They wanted a deejay to work for a buck an hour from above a condemned bus station in Wautaga County, the second poorest county in the country."

A few months later, White returned to California and landed his next radio job because "a guy had a heart attack and they needed someone right away." White next went to Armed Forces Radio in Alaska, and it was there that he ended up landing a noon anchoring job on TV. He says, "The news director went back east on a fellowship and never came back, and I became an instant news director. I was a one-man news department. I did everything, plus manage an FM station which we put on the air while I was there. I learned by doing and loved it."

THE SPORTS TRACK

Over the years, the reporters and anchors I've met in local TV's sports departments think they've died and gone to heaven. They love sports more than life itself, and now they are being paid to report on their passion, to go to sporting events, including all the biggies—the World Series, Super Bowl, maybe even the Olympics. On top of that, they don't even have to buy the tickets! They're being paid to interview the hottest sports stars, athletes that others only dream of meeting.

With only a handful of sports anchors, reporters and producers, even in the largest stations, competition for these positions is unreal. Those interested in sports reporting and anchoring may be able to get some experience in college or even in high school. Remember Michael Scannell's enterprising story in chapter 5? Many aspiring sports reporters hustle for college internships in local radio and TV sports departments. Many who end up in TV sports learn the ropes on the radio side.

My favorite Cinderella story in sports is that of Vernon Glenn, KRON's weekend sports anchor. He was a 10th grader in Virginia when he realized he wanted to be a sportscaster, but after graduating from the University of Virginia in 1984 with a broadcast journalism degree in hand, he couldn't find a job. He says, "It was catch-22. Nobody's going to listen to you if you don't have experience. Well, how do you get experience if you can't get in? Out of necessity, I took a sales job with Reynolds Aluminum Company. Every day I was there, I would pick up the phone and call local radio and TV stations just to find out who would listen to me. Nobody would, then one Sunday night, I was listening to a sports talk show called 'Let's Talk Sports with Chuck Noe,' the former coach of South Carolina and of Virginia Commonwealth University. I called him up after the show, I got through and I said, 'I want to be a sportscaster. I hate my job. How in the world do I get in? Nobody'll listen to me. If I can't get in the business, at least I'd like to be around it. So would you mind if I came down and sat in on the show, see what you do?' And the guy said, 'OK, young fellow, come in.'

"So I came on in that Sunday, had a great time. He took calls. He even gave me a mention on the air, which was really a thrill. And then I came down the following Sunday and the following Sunday and he took a liking to me so he made a call to a guy he knew at this AM daytime station. He was just looking for someone to read scoring updates at 8:00 a.m. and 5:00 p.m.

"I got the job. It was for free. I'd go by at 8:00 a.m., do my little rip and read, then go to Reynolds Aluminum to my regular job, call in at 5:00 and do a phoner, after having called a sports update wire from the office to get my information. So I'd do this day after day after day.

"I started going to ball games with my little tape recorder, practicing my play-by-play, and I met the sports director for the ABC affiliate in Richmond. I gave him the same spiel; I said, 'Boy, if I can't be in the business, I'd like to at least be around it. Would you mind if I came out at night and kind of helped you with your late show, and kind of looked over your shoulder? I could be an extra body for you.' Because Richmond was a nonunion town and you shot, you edited, you set up your own live shot, it was a one-man band situation, so at least I was there, another body, so he could get up and go get something to eat and come back.

"A few months later, the weekend sports guy quit, so they were looking for someone they could pay cheap, 25 hours a week, to come in and do the weekends. I saw all these resume tapes coming in from smaller markets, and they were pretty polished but it's not something that I couldn't do. I talked the powers that be into letting me do a little demo after the early show. Did another one the next day.

"Got the job on a Thursday, quit Reynolds on Friday and my first day ever on the air was that Saturday. Took an extra pair of pants just in case I messed up that pair."

It wasn't a cake-walk for Vernon, however. He received hate

calls for several months "because I didn't know what I was doing." And he took a pay cut of $12,000—from $22,000 at Reynolds to $10,000 at the station. But in a relatively short time, it had paid off.

His second year, his salary was upped to $13,000, and his third, to $17,000. Then he signed with an agent. It was September of 1987, just over three years since his graduation. He says, "The next thing I know, I'm at WBAL, the CBS [now NBC] affiliate in Baltimore, making three times as much money. Things really started taking off. I thought I'd be in Baltimore for a very long time, but then in January of 1990, I flew out to KRON for an interview. And I got an offer on my 28th birthday. I signed and I've been here ever since."

Vernon is a perfect example of someone who created his own contacts. He says, "It takes perseverance. That innate thing of always having hope that at least you have a shot at your dream, and when you get that shot, at least try not to blow it. I knew that being a sportscaster was something I always wanted to do and I'm doing it now."

So You Want to be Oprah?

As difficult as it is to achieve, the route to an on-air news job is considerably more direct than the route to other on-air jobs, such as talk and game show hosts, or their side-kicks.

For an on-air personality outside of news, I would recommend a broad liberal arts education, including some speech classes. Those who succeed share a combination of looks, intelligence, a sense of humor, natural curiosity, an ability to listen and conduct an interesting interview. They also have guts and luck. Many have worked in other areas of media, theater, comedy or journalism.

Although talk shows on local network-affiliated stations may be disappearing, there are many other opportunities in cable and syndicated programming. Most likely, you'll need to go the route of the small market to gain some experience and compile a resume tape.

One friend of mine, Ruby Petersen Unger, currently producing shows on how to talk to kids about sex and drugs, got her first on-air job, amazingly, through the *San Francisco Chronicle* classifieds, as Ms. Nancy, host of "Romper Room" at KTVU in Oakland. Ruby had majored in broadcasting at the University of Wisconsin and later, after a marriage and career in retailing, went back to Marquette University for training as a speech therapist for public schools. She fell in love with the Bay Area while on vacation and moved to San Francisco, getting a job with the March of Dimes Walkathon. It wasn't long before she hit the open mikes, doing stand-up comedy at the Old Boarding House and the Holy City Zoo. Then she landed a job as a comic for a male cabaret on Broadway in San Francisco. (Actually, it was one of the first male strip joints and I personally found her and the whole scene screamingly funny.) After the show ended, Ruby spotted a classified ad under "teachers" for "Romper Room" on KTVU. It had everything she wanted—kids, comedy and broadcasting. Hired in February of 1981, she went to Romper U. in Baltimore to learn how to be Ms. Nancy (leaving the male strippers behind, if you'll pardon the pun).

Ruby got very involved in AWRT, serving in just about every capacity, including president, and meeting people who helped her career tremendously. One contact through AWRT was Charlie Seraphin, KCBS Radio program director. After she subbed for one of their talk show hosts, Seraphin talked Ruby into doing a series of late night comedy radio shows live from the Fairmont Hotel. It was your "San Francisco Saturday Night Comedy College of Knowledge with Professor Ruby Petersen"—two hours of live comedy featuring a band, Swing Fever, and comedians.

Another gig was as deejay for KLOK-FM in the middle of the night, then she was tapped to be a regular panelist on a local game show called "Claim to Fame," in the vein of "What's My Line?". Although the show consistently won it's time period in the ratings, it was canceled after about a year due to a legal dispute over its origins.

Ruby subbed for the regular hosts on Bay Area talk shows, and then was chosen to host a fascinating program on a San Francisco independent station called "KOFY Personals." Gutsy singles appeared on camera to let the world know they were available. Interested viewers called a (900) number to leave messages to offer themselves up as potential dates. This was Ruby's "video Yenta" period.

She says, "If you get your toe in, things will take off." Ruby's had a successful, if checkered, career, which I attribute in part to her wonderful sense of humor. She's naturally a very funny person.

Ruby gives credit for most of her jobs to the friends and contacts she met while donating her time to AWRT. She swears by professional associations for contacts and career enhancement.

I've known a lot of talk show hosts and radio personalities, and have met a lot of network series stars. Most have loved their work, when they have it. But they're an anxious lot, never knowing when the gig will be up. And there's the rub. It requires tremendous fortitude to deal with the fickle, ever-changing marketplace. You can spend months or even years getting that perfect job, only to have the show canceled two months later. Then you're out pounding the pavement again. And, particularly if you're a woman, you're working to maintain your figure and youthful appearance as the clock is ticking.

I say, go for it if you've got to have it. But don't give up your day job.

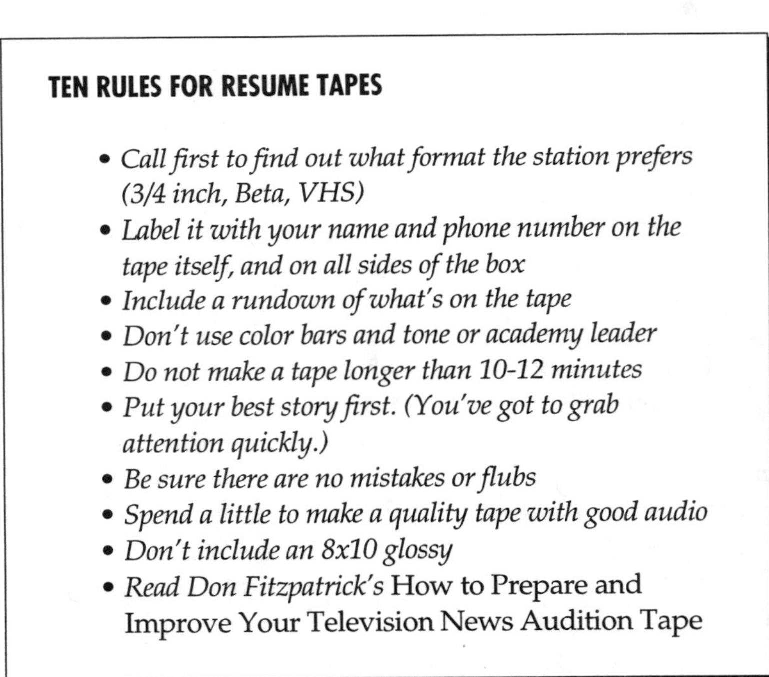

TEN RULES FOR RESUME TAPES

- *Call first to find out what format the station prefers (3/4 inch, Beta, VHS)*
- *Label it with your name and phone number on the tape itself, and on all sides of the box*
- *Include a rundown of what's on the tape*
- *Don't use color bars and tone or academy leader*
- *Do not make a tape longer than 10-12 minutes*
- *Put your best story first. (You've got to grab attention quickly.)*
- *Be sure there are no mistakes or flubs*
- *Spend a little to make a quality tape with good audio*
- *Don't include an 8x10 glossy*
- *Read Don Fitzpatrick's* How to Prepare and Improve Your Television News Audition Tape

Resume Tapes: The On-Air Talent's Calling Card

Although you won't be breaking into TV at the on-air level, if being on camera is your ultimate goal, you need to know about resume or audition tapes. And the sooner the better.

If a picture is worth a thousand words, a resume tape is worth a whole lot more. The boss has to see you in action on videotape to know if you're the one. And don't even bother to send that 8x10 glossy with or in lieu of a resume tape. It says nothing about how you'll look on camera and will end up in the circular file, or worse, as Don Fitzpatrick suggests, as a doodle pad for the news director. The Hollywood glamour photos are only sent by naive applicants or those who are particularly enamored of their looks.

So we're back to the old catch-22. How do you put together a resume tape without on-air experience?

GETTING STORIES FOR YOUR TAPE

Perhaps you did some reporting or anchoring at your college station. That's a start. If you've gotten a job as an intern or news assistant, you might be able to talk a photographer into taping you in the field, even if it isn't going to make air.

Wendy Burch suggests this strategy for getting material for your tape. "You go to the assignment editor who's pressed for time one day and you say, 'I'll go get the sound bite you need.' You go out, you buy the photographer lunch. He shoots a standup for you and you ask him to shoot just a little extra video so you can put together a package."

The assignment editor gets his sound bite, and you get what you need. Wendy says that's how she put a tape together while interning in Salt Lake City. She also got lucky because her station instituted 24-hour news updates. She explains, "Sometimes the main anchors got tired of doing the updates, so at 11:59 at night, you could catch me on the air. It wasn't much, but I got a little bit of anchor experience."

Professor Hewitt warns, "Think about how you'll look to the people getting your tapes. Even though you're neat and respectable looking at the interview, if you were a long-haired, wild-eyed freak on your tape, it'll look bad. So when you're doing school projects, think ahead. Don't be afraid to spend a little money making a good tape, to have good quality, good audio. Remember, the three to five stories you did in college will be competing with others who did three to five stories a week and have lots from which to choose."

A resume tape composed of college reporting and rewritten, reedited stories that never made air will not get you a job except in the smallest of markets. And you shouldn't try to pass it off as something it isn't. The tape won't compete well with those from experienced reporters, but it will give a news director an idea of how you come across on camera. It also indicates a level of desire and initiative on your part.

WHAT TO INCLUDE

At a 1992 regional RTNDA convention, Don Fitzpatrick said, "News directors are looking for people who sell themselves." He screened a resume tape for the audience that particularly impressed him and then explained the appeal, "Sally sold the hell out of every one of those stories. She looked you right in the eye, and she told you why that intersection was important. She made that story compelling."

News Director Doug McKnight says, "The story that stands out for me is the one that you enterprised. I'd like to see a story that kept me interested all the way through. Anyone can do a highlight tape." Also keep in mind that news directors are looking for good writing.

Joe Fragola at BayTV has been looking at a lot of tapes lately, and he says, "Give me a slate for ten seconds, go to black, boom, come up with your first piece. I would go with a montage of anchoring, reporting, reporter stand-ups. Have that montage for a minute, a minute-15 max. Ten seconds here, ten seconds there. Then get into your package pieces. So they see who you are."

Wendy Burch says, "It's really important to make a good impression. Be sure not to bore the person watching those tapes. I like to see you right off the top. Video montage at the top. It showed my news director that I had done more than just three stand-ups. It gave him a chance to see how I anchored, how I looked in the field, and what I looked and sounded like. It's only two minutes long, but I'd probably make it about half that long if I had it to do again. It's just a tease of what's to come. Then get into your packages.

"If the job listed is for anchor/reporter, be sure to put both on. The whole tape should only be 10-12 minutes, concise enough that a news director can watch the entire thing. Three packages are the key. First one a strong story; I would suggest a crime story. It would be great if it was a spot news story. That will kill two birds with one stone. It will show you can get the facts and do it on a deadline. The second story I would recommend would be an issue-oriented story, and finally a feature story, perhaps about someone unique in your area, not the standard pumpkin patch piece."

Rosemary Wesela advises, "Don't send a resume tape with a mistake, a flub, or with anything on it where you say, 'Oh, I'll explain that when I get in to see him or I'll let him know it wasn't my best hair day.' If you can't be absolutely proud of what you're sending, don't send it. Also, don't send a resume tape where you have 15 standups in the beginning. He does want to know what you look like, but he wants to hear how you write, he wants to know if you can focus a story, he wants to know how you do the package. A couple of standups will do."

If you're putting together a tape, by all means, get Don Fitzpatrick's guide *How to Prepare and Improve Your Television News Audition Tape.* It'll give you your best shot at producing a tape that news directors will actually watch for more than five seconds. Don tells you how to grab their interest in the first few seconds, what to put first, depending upon the position for which you're applying, and precisely what to include.

His advice is based on the thousands of resume tapes he's seen, on the feedback he's received from his clients over the years, and from the results of a questionnaire he sent to more than 300 news directors and talent search folks at the big TV consulting firms. He includes specific advice for reporters, anchors, weather and sports reporters, talk show hosts and producers.

He not only talks about what to include, what to leave out and specifics on the order of stories and length, but he also includes sections on resumes and cover letters, pet peeves, how to ship the tape to news directors and follow-up. In the back of the guide is a talent fact sheet which should accompany the tape you send to him. It asks for details about your background, your market or geographic preferences, salary sought, etc.

In Don's second booklet *How, When & Should You Hire a Talent Agent to Represent You?*, Don offers all the pros and cons of working with agents, how to select one, questions to ask them, and even a five-page partial listing of agents, their specialties, fees, addresses and phone numbers.

If you're considering an on-air career—or even if you're already on the air—these guides are the best little investment you can make. Each guide costs $15.00. If you want both, the cost is $25.00. Send your check or money order to:

Don Fitzpatrick Associates
408 Columbus Avenue, Suite 1
San Francisco, CA 94133
(415) 954-0700

IDENTIFYING YOUR TAPE

Advice offered by everyone I've spoken with: call first to find out which format is preferred—3/4", VHS or Beta. (Some prefer 3/4" although VHS costs less to produce.) Don't use color bars, tone or academy leader, just a slate with your name and phone number. Include a rundown of what's on the tape but don't attach it to the tape because it can't be read from inside the machine. Be sure the tape box and tape itself are labeled all around, so they can be easily identified even if they're in a giant stack of tapes.

WHEN TO SEND TAPES

To send or not to send? That is the question. Sending out a hundred or so tapes can get very expensive. News directors won't take the time to return them. (You might get some back if you enclose a self-addressed envelope.) Some say go ahead and send a tape, that a news director always wants to see it. Others advise not to send unsolicited tapes because you'll be wasting your limited resources. If there's no job open, the news director will have other priorities and won't look at it. Since phone contact is so important in your job search and each news director has his or her own style when it comes to hiring and interviewing, I recommend calling first and asking the news director or assistant what s/he would prefer.

Rosemary Wesela agrees, "Every station handles job applicants their own way, so call first. If you send a resume and tape and there is no job, they might not look at it at all, or they might look at it in six months. When you call, ask if they have any openings and find out how quickly they plan to fill them. Find out how they operate. Again, act like a reporter trying to get information. Focus on the story, which is, 'How do I get in?' You need the secretary or assignment desk person on your side. They will tell the news director, 'This person listened, he paid attention.'"

FOLLOWING UP

This is picky, but Rosemary doesn't like job-seekers to call and ask if their tape arrived. She *does* think it's appropriate, however, to wait two weeks, then call and ask, "Did you get a chance to look at my tape?" She adds, "Don't tell me you don't want to send your tape because you want to be here when the news director looks at it. The news director does not want you there when he's looking at your tape. If you're sitting there, he's going to have to look at the whole thing and he doesn't want to. If he's interested, he'll call you so fast you won't know what to do. If he's interested, he won't put it back on the shelf and let it sit. He'll call you right away to say, 'Come on in. I want to see you. I want to see more of your work.'"

THE NEWS DIRECTORS' SHOPPING CHANNEL

UPS, Federal Express and the U.S. Postal Service are getting rich off the shipping of resume tapes to news directors around the country. Most news directors have, at any given time, a stack of tapes from reporters and anchors hoping to improve their lot.

Many news directors simplify their lives by relying on the services of Don Fitzpatrick Associates, the premier headhunters in the country for news talent. On a regular basis, Don, who used to be CBS' head headhunter, tapes all the newscasts in roughly the top 200 markets (leaving out only a handful of the smallest markets). When a news director is looking for an anchor, s/he can ask Don for resume tapes and backgrounds of, say, the ten best African-American females with at least three years' experience.

The news director may also look at tapes sent directly by talent via the mail or courier, but Don's comprehensive tape library can target candidates who meet specific criteria. Since Don has current tapes of 15,000 anchors, reporters, show hosts and producers, the news director can be confident that the perfect person for the job hasn't been overlooked. Often the news director wants to hire someone who is enjoying success and popularity in another market (or even the same market), who isn't even looking for a job.

Don also incorporates unsolicited tapes into his library. Here's how he works. "We work for and are paid by networks, television stations, syndicators and so on. We get about 40 tapes a day, unsolicited, into our offices. Each one is 12 to 15 minutes long and, unlike a news director who will pop the tape into the machine and give you 3-10 seconds to make the sale, we watch the tape from start to finish, all the way through. We log in all the material and then we write comments and notes to ourselves. A lot of times, the best story on the tape might be the third story, even though you thought it was the first. We reedit the tapes in our office before we send them out."

Once, Don got a request for reporter candidates from a general manager he knew to be an avid marathoner. The GM was an impatient guy who would rarely watch more than a few seconds of a resume tape. Don had received a reporter's tape that included a story on a marathoner, but it was the fourth story on the tape. Don says, "So we reedited it, putting the marathon story first. The GM got hooked on the first story and ended up watching the whole tape and hiring the person."

(In the case of a producer's tape, Don Fitzpatrick Associates keeps the tape of the entire show intact, since a major element in evaluating a producer is how s/he put the entire show together. But this chapter is not about producers' resume tapes. If you're interested, refer to Don's booklet on tapes.)

About shelf life, Don says, "If you send us a tape, we'll look at it all the way through. We'll put it on the shelf. We'll keep it for about a year. If you update the tape, we'll take your old tape off and replace it with the new, and if you don't replace it, we'll toss it after a year because we don't know what's happened to you. Yes, most of our clients are clients like CBS, CNN, ESPN, but we also work for stations in Yakima and Klamath Falls, Medford, Oregon, Salinas/Monterey. We're in large markets, but we're in smaller markets as well."

Don't Forget Your Cable

23

Careers in Local Cable

The growth of cable television in the past two decades has opened up a whole new universe of career opportunities, from local cable systems to multiple system operators (MSOs) to cable networks.

Prior to cable, broadcast TV was an only child, and a spoiled one at that. With such limited competition, the three networks and local TV stations made tremendous profits. It was commonly suggested that TV stations and networks had a "license to print money." I look back fondly on those years of ample budgets and annual 10% raises.

Then came cable and competition. Big Brother broadcasting had to share the advertising pie. Today, cable advertising brings in billions of dollars and, although it hasn't yet caught up with broadcasting revenues, cable revenues are increasing at a much greater rate.

With cable now reaching into the majority of American homes and with satellites beaming cable news and programming to the entire world, cable television has come of age. In fact, the Cable News Network, which wasn't even launched until 1980, is now one of the primary international news sources in the world. CNN not only brought the Gulf War into our living rooms, but you can be sure that government officials from all over the world were watching the conflict unfold, in real time.

LOCAL CABLE SYSTEMS

The number of local cable systems around the country is growing every day. Suffice it to say that there are well over 11,000 local systems, supported by subscriber fees and advertising.

You'll find opportunities in the technical area, in installation and maintenance of the cable and in problem-shooting in the customer's home. The entry level tech job is that of installer, who needs some trade school background and mechanical savvy.

The administrative department coordinates the work of the technical department and handles billing, other accounting functions and customer service. Entry level positions include those of customer service reps and service dispatchers.

The marketing and advertising folks are bringing in the revenue. Sales reps, who generally need some previous sales experience, sell basic cable and premium services to subscribers, sometimes via telemarketing. The advertising sales team sells commercial time to local merchants.

In the area of programming, the local cable system management makes decisions about which channels to offer their subscribers. They choose from local TV affiliates, independent stations, and the current selection of more than 100 national and regional cable programming networks. Approximately half also have their own LO departments. Many of these full-blown production operations produce a volume of local programming that exceeds anything a local broadcaster attempts.

OPPORTUNITIES IN LOCAL ORIGINATION

As local programming departments at broadcast stations have been sliced, diced or totally obliterated, many cable operators have been expanding the output of their LO departments. It's not unusual to find cable operators producing several hours of original programming each week.

Many of them cover local high school sports, live or on tape. Several feature a high school football game of the week. Some cover basketball, softball and even lacrosse. LO departments cover local festivals, parades, county government, community events, and local entertainment such as school or church choral performances. Several produce children's programming. Some have live call-in shows for sports enthusiasts. Other call-in shows are forums for local politics, medicine and legal concerns. Some cable operators offer programs geared to minority issues, some have news magazines, and some are even doing local newscasts. A few produce documentaries and educational programs. Suburban Cablevision in northern New Jersey has one community program that involves all 42 mayors in the area.

Barrett Lester, program manager for Continental Cablevision of Lawrence, Massachusetts, suggests that there are a couple of reasons for the growth of LO. The cost of equipment has been decreasing. Video cameras cost considerably less than they did ten years ago. Also, with added competition for cable installation from the telecommunications industry in the not-too-distant future, cable operators realize that they'll have the edge if they've established a tradition of service to the community through local programming.

Whatever the reason, the growth of LO means lots of opportunity in cable. LO departments offer not only internships for students at nearby colleges, but also volunteer opportunities. Many are very flexible with their volunteers and interns, working around their schedules if they're students or employed elsewhere full-time. Remember that interns and volunteers have the edge when paying positions open up.

John James, program manager for Syracuse NewChannels in New York, says, "We're always looking for talented young people. We put them on part-time crew positions after their internships. This gives them a wonderful opportunity to network, as there are two cable companies and four affiliates in our market."

ACCESS

Cable operators' franchise agreements usually require that they provide local access programming. Unlike local origination, local access is programming produced by community people, not the cable operator's professional staff. A fun example of access programming is the fictional spoof, "Wayne's World." Sometimes access shares the same channel with LO, but occasionally a second channel is allotted just to accommodate access programming.

Due to the access clause, many local cable systems offer training to individuals who want to learn video production. The classes, which cost a nominal amount if they aren't absolutely free, focus on the use of the video cameras and editing equipment. Individuals may submit program proposals for the access channel and can even sign up to use the equipment.

Production in LO and access involves the same skills. Studio productions involve producers, program talent, audio and lighting technicians, camera operators, floor managers, the link between the control room crew and the talent in the studio, studio technicians and video editors. Volunteers and interns can learn a great deal about production in local cable.

Local cable systems vary greatly in their LO output and so will the opportunities. They're changing and expanding as we speak, so don't make assumptions based on market size or what you've observed in the past. Check them out.

MSOs

MSOs are businesses that own and operate from a few to a few hundred local cable systems. The personnel at an MSO headquarters or regional offices provide support services to their systems. They may serve as central purchasing agents for the systems, handle budgets, maintain FCC records, and look for new cable ventures.

The departments you can expect to find at MSOs are corporate engineering, operations management, sales and marketing, public affairs, human resources, finance and legal. MSOs, however, do not produce programming.

Cable Networks

If you enjoy A&E, ESPN, Lifetime, MTV and CNN, a career with a cable network may be fairly appealing. However, I must warn you that breaking into a cable network is right up there in difficulty with breaking into a broadcast network. Also, similar to broadcast, programming for many of the cable networks is produced by outside production companies.

Cable programming networks fall into three categories. Basic cable networks are primarily advertiser-supported and available with the monthly cable fee you pay your local cable service. Pay-TV networks, usually movie or sports channels, often provide programming without commercials and are available for a fee as an addition to your basic cable offerings. Pay-per-view is event programming, often sports or movies, which can be bought on an individual basis.

The *Hollywood Reporter* regularly publishes production information that includes names and addresses of the cable networks and independent production companies that supply them with product. The *Broadcasting & Cable Yearbook,* available at your local library, also provides addresses and phone numbers of the cable networks. If you call concerning job or intern opportunities, you're usually connected to a recorded message telling you that you can mail in your resume and cover letter. Many, however, don't give out information about job openings, as they like to promote from within. On the other hand, most do advertise some of their job openings in the broadcasting and cable trade publications.

INTERNSHIPS AND TEMP POSITIONS

MTV, ESPN, Lifetime, USA, the Discovery and Learning Channels, and CNN and its bureaus do have internships for students. Most are unpaid, although the Discovery and Learning Channels pay their interns $7 an hour. Discovery also actively recruits interns from the communications departments of local colleges and universities near their headquarters in Bethesda, Maryland, including the University of Maryland, American University, Howard, Georgetown and George Washington. Read about CNN's internship program later in this chapter.

As is true throughout the entire business community, temporary jobs at the cable networks often turn into permanent jobs. A number of cable staffers told me that they were promoted out of their original temp positions to other jobs, or that their temp jobs became permanent.

MINORITY RECRUITMENT

The cable networks banded together more than a decade ago to offer opportunities to ethnic minorities through the Walter Kaitz Foundation. The national nonprofit foundation, fully funded by the cable industry itself, recruits ethnic minorities with college degrees for management, professional and technical positions nationwide from outside the industry. Although it does look at students for technical internships, it primarily seeks people with strong transferable skills and three years of professional experience in another industry.

Write for information on its programs and fellowships.

Walter Kaitz Foundation
660 13th Street, Suite 200
Oakland, CA 94612
(510) 451-9000

ASSOCIATIONS

Networking is also the name of the game in cable. Magazines like *Cable World* run monthly columns listing key people and their job changes. In addition to using the broadcasting/cable trades as a resource, the NCTA, the chief trade association in the industry, meets in May. Whether or not you take advantage of the educational and networking opportunities at their convention, do write for their publication called *Careers in Cable Television*. In addition to describing in detail the operation of local cable systems, MSOs and cable networks, the booklet gives addresses and phone numbers for the 50 largest MSOs, the cable programming networks and cable professional associations and a bibliography of cable TV publications.

National Cable Television Association Inc.
1724 Massachusetts Ave., NW
Washington, DC 20036
(202) 775-3550

Another great resource for information and networking is your local cable organization. (Check with a local cable operator.) Also, you may have a chapter of the National Association of Minorities in Cable (NAMIC) in your area.

CNN

When CNN was new on the scene in the early '80s, those of us who hosted career panels used to suggest it as a great place to gain experience. After all, with 24 hours a day devoted to news, we considered CNN the marathon of TV news gathering. It was trial by fire for CNN's newswriters, producers, shooters, editors, reporters and anchors.

In those days, we thought of CNN as a stepping-stone to the big time. Today, with its international reputation and viewership, it's no longer a a means to an end. For many, it's the final destination. As network news operations have shrunk, CNN has expanded. Today CNN employs some 1,700 staffers in Atlanta. Their worldwide employee count is around 2,500. CNN's domestic bureaus, besides headquarters in Atlanta, are in New York, Los Angeles, Chicago, Washington, DC, Dallas, Miami, Detroit and San Francisco. Their international bureaus number 20—in Amman, Bangkok, Beijing, Berlin, Brussels, Cairo, Jerusalem, Johannesburg, London, Manila, Mexico City, Moscow, Nairobi, New Delhi, Paris, Rio de Janeiro, Rome, Santiago, Seoul and Tokyo.

CNN warrants a section of its own for several reasons. CNN is unique. Being the round-the-clock news operation it is, it offers terrific opportunities for those with a passion for news and production, promotion from within and excellent training for its employees. And without exception, every CNN employee with whom I have spoken is very enthusiastic about the company.

CNN's Internship Program

CNN internships provide excellent opportunities for budding journalists to learn the trade. If you're interested in interning at one of CNN's bureaus, contact the bureau directly or a CNN representative in New York City, Washington, DC, or London. CNN will provide you with a list of contacts and addresses for all their bureaus as well as the contacts in New York, Washington, DC and London.

If you're studying in the Atlanta area, or would like to spend a summer there, Turner Broadcasting System, Inc., offers a wide variety of internships at all their divisions, including CNN. In Atlanta, internships are offered quarterly to college juniors, seniors and graduate students. Although the nonpaid internships are designed for student volunteers who are able to receive academic credit from their university or college, CNN does not require that you receive credit.

When you apply, you're asked to indicate which departments within the Turner Broadcasting System interest you. Some of these departments are in corporate, some with CNN, others CNN International, others Headline News, TNT, and their Sports or SuperStation. Internships may not be available every quarter in all of these departments, and, at the same time, are not limited to these departments:

- *Advertising*
- *Business News*
- *CNN International*
- *Earth Matters*
- *Entertainment*
- *Environmental Unit*
- *Features*
- *Future Watch*
- *Graphics*
- *Headline News*
- *International Assignment Desk*
- *Marketing*
- *Medical News*
- *Network Earth*
- *Nutrition*
- *Photo & Video Services*
- *Political Unit*
- *Promotion*
- *Public Affairs*
- *Public Relations*
- *Sales*
- *Science*
- *Spanish News*
- *Special Assignments*
- *Special Reports*
- *Sports*
- *Sports Marketing*
- *Travel Guide*
- *World Report*

If you're in school outside the Atlanta community, you can apply for the summer quarter internship. TBS will even provide you a partial list of short-term housing options in the area. They'll also work around your schedule and allow you to intern up to 40 hours a week.

Jacqueline Trube, internship coordinator for TBS, says, "An internship with TBS provides valuable on-the-job experience, a chance to make future business contacts, and the opportunity to obtain an important recommendation for future employment." Of course, an intern's experience at CNN will depend largely on the person's initiative, the department worked in and his or her supervisor.

To be considered for an internship at CNN, you'll need to fill in their application and send a current resume and cover letter describing your professional areas of interest, your experience and how you would benefit from the internship. You'll also need two letters of recommendation (one from a faculty sponsor), a letter concerning course credit, if applicable, and an official college transcript. Write for an application to:

Internship Coordinator
TBS, Inc.
One CNN Center
Box 105366
Atlanta, GA 30348-5366

CNN'S VIDEO JOURNALIST PROGRAM

CNN has a philosophy of training its own and promoting from within. The majority of their writers, producers, field producers, techs, editors, directors, TDs, master control personnel, associate producers, supervisors, desk assistants and assignment editors at CNN and CNN Headline News started with the company as video journalists.

VJs, as they're called, do not, however, advance to on-air positions. Most CNN anchors and correspondents are hired after five to eight years experience at a local station.

The VJ position is the entry level position at CNN and CNN Headline News and has two tracks: technical and editorial.

If you're interested in the technical side and CNN's interested in you, you'll be asked for references and a list of the equipment you've operated. If you're headed down the editorial track, you'll be sent a writing test to take at home. The test is also administered at the Atlanta headquarters if you're in the area or, occasionally, at one of the CNN bureaus. You're given three or four wire stories from which you're asked to write on-air scripts 20 to 25 seconds in length—in one hour.

All VJs begin in Atlanta at an annual salary of $17,500 based on a 40-hour week. They comprise the floor crew for the CNN and Headline News broadcasts. At CNN, all VJs begin working as floor manager, studio camera operator, script ripper, TelePrompTer operator and news playback operator. After six to twelve months, they head off on the tech or editorial track. For both tracks, there's a natural progression of promotion, although promotion is not automatic. It's based on merit and is decided by an evaluation process that includes testing.

Those on the editorial track proceed to tapes production coordinator, then to production assistant, then to associate producer. From there, advancement is either to line producer, tapes producer or writer/producer-trainee to producer.

At CNN Headline News, the advancement for editorial VJs is to editorial assistant, which involves ripping scripts and training as a writer, to writer, associate producer, copy editor, producer and supervising producer.

On the editorial side, you'll work with a coach or senior writer who will help you improve your writing. You'll write for air, and the trainer will correct and edit your work. The coaches are simply more experienced writers who have been at CNN for several years. Suffice it to say that the VJ program provides a thorough understanding of the entire operation.

The VJ technical track offers two options. The first option is CNN news playback VJ to news playback/dubs for CNN International to international technician and library management system operator to CNN audio to master control to technical director to on-air director; and there's a variation on this option that encompasses electronic graphics. The second option is news playback VJ to feeds operator to videotape editor to South East Bureau sound tech rotation to field photographer or CMX editor.

The VJ program has no time limit. Your rate of advancement will depend upon two factors—your ability and job openings on your track. Because CNN promotes from within, promotions have a domino effect. If one person high up the chain leaves or gets promoted, opportunities are created all the way down the line.

One small downside of promotion, which is common throughout the business, is that each time you're promoted to the next level, you pull the graveyard shift of 1:00 a.m. to 10:00 a.m. until you gain some seniority in that position. This program is not for wusses.

On the editorial side, CNN is looking for people who can deal well with pressure, who can write 30 scripts in a day, eight hours a day. They prefer applicants with degrees from four-year universities or colleges in either journalism/communications or liberal arts. They also look for people with additional TV production and studio experience, usually in the form of one or more internships in newsrooms and/or studios outside of the college environment. Editorial experience from school newspapers or other print media is acceptable. Finally, and this should be obvious, they want candidates who are interested in news, people with a daily diet of reading and watching news.

If you're in school, one of the best ways to get into the VJ program is to land an internship in Atlanta or at one of CNN's bureaus. If you're interested in CNN, you'll want to get in as early in your career as you can because they'll get you on the VJ track and train you themselves. CNN training and experience will serve you well no matter where your career takes you.

The Big House

The Networks

Yes, the networks have a very glamorous aura about them. They provide a common cultural bond that stretches from coast to coast and beyond. Many of us grew up on the big three with some PBS thrown in for variety. For a couple of decades, before our options started multiplying, the networks were synonymous with TV.

Increasing competition and other economic influences of the '80s and '90s have affected the networks, like so much of the business world. They've had to tighten their belts, and managers in all facets of network TV are working to make smaller budgets stretch further. At the same time, they're looking to the future, discovering new ways to capture a fragmented audience. They're developing their own cable outlets, getting involved in multimedia and paying close attention to technological advances.

Network news operations closed most of their domestic and many of their international news bureaus in the '80s, and have cut back in employees across the board. You'll find almost all network employees for ABC, CBS, NBC and Fox in two cities: New York, which is center for news operations and news-oriented series, and Los Angeles, whence emanates most of the entertainment fare, although a very small percentage is produced by the networks themselves.

PAGE PROGRAMS

While the traditional foot-in-the-door job at a local station has been in the mailroom, at the network, the entry level job is the proverbial page. The page programs vary from network to network and from coast to coast, depending on who's running the program, the network's production needs and the rules and traditions that have been established over time.

One of the primary functions of pages is to work with the audiences who come to the tapings of shows. At NBC in New York, pages corral the guests for the daily taping of "Donahue" and "Saturday Night Live." At CBS in New York, they do the same for "Late Show with David Letterman." Pages on both coasts also give tours of the network facilities.

NBC's page program in New York is a 10-month paid program which requires a degree from a four-year college. ABC in Los Angeles, on the other hand, has only two requirements: applicants must be U.S. citizens and 18 years of age or older.

If I were starting over, I'd definitely head to L.A. to take advantage of the ABC program. First of all, L.A. is the heart of the entertainment industry. Not only is it Production Central for TV programming, but it's also the filmmaking capital of the world. As a result, the pages at the L.A. network studios have a lot more contact with people in the business and more opportunities to move around the movers and shakers, not to mention the stars. For example, at the Academy Awards each year, an ABC page is assigned to each of the major presenters to assure that they're in the right place at the right time and that their needs are met.

The ABC page program has been run for the past 24 years by Joseph DiSante, Manager of West Coast Administration for ABC Television. The page program, just one of his responsibilities, is near and dear to his heart. As of the end of 1994, he says some 275 of his pages have moved up in broadcasting. Some have won

Emmys. Many, he admits, have surpassed him in the industry. Dozens of them are still working for KABC, the network-owned station in L.A. that shares facilities with the network. Connie Borge Youngblood, KABC's current programming director, was one of his best pages a few years back. He says the way she worked her was up is a real "star" story.

DiSante's pages, which currently number 19 (although in the past he's had as many as 100), are similar to a temp staff assigned out as needed to programs, news and production houses producing ABC shows. In addition to working on shows, conducting tours, coordinating audiences and working with the talent, they're assigned to all kinds of offices, to plant services where the sets are constructed, and to help out throughout the KABC operation.

DiSante humbly admits that ABC has "absolutely the best page program. The ABC page staff has a wonderful reputation. A producer or director knows they're getting quality, first-rate people. We've gone into situations where there's total chaos, and everyone breathes a sign of relief when the ABC pages walk in."

ABC pages are permanent part-time staffers, starting at $8.26 an hour. After six months, they get a raise, and not long after that, they reach the third and highest tier at $9.50 an hour. They're also paid overtime after eight hours and receive meal penalties (are paid extra if they work through a meal). If asked to travel, they get travel pay. The program is open-ended. They can stay on as pages as long as they'd like.

Many start as pages while in college. DiSante works around their school schedules and also assigns them according to their interests and skills. He says, "If someone's interested in writing, they're not going to be happy in the accounting department. It's a great part-time job. I've put a lot of kids through college."

DiSante has a keen instinct for the kind of person he wants in the program. And he's more than happy to go to college campuses around the country to speak on this very subject. He's looking for people who are hungry to get into the business. He says, "I know when a winner walks through the door. My favorite question is 'Where do you see yourself five years from now?' If they say, 'Running the company,' I'm not interested. I like people with a realistic approach to the industry. I like people who aren't sure where they belong in the industry but want to find out. As a page, you come in on the ground floor and you're a part of and experience all aspects of the industry from engineering, accounting, human resources, camera work, writing. It's a dues-paying process. You're not going to become the head of daytime programming in a year. It's a tough, tough industry. You have to start at the bottom and work your way up."

DiSante hires whenever he has a need for additional pages. He's always looking at resumes and cover letters. He says he can see from an applicant's volunteer activities in the industry whether he or she would be a good candidate. He likes to see volunteer jobs on a resume because it proves that the applicant is not just interested in the money. He wants people who appreciate the beauty and the dynamics of the industry. He says, "If you're good, the money will be there. If you get good at it, you *will* make money."

DiSante's advice for people who want to break into the business: "If you're production-oriented, there's only one place for it, and that's L.A. If you're interested in production, this is it. For people who are interested in news and news journalism, my advice is to leave this market. If you want to be an on-camera reporter, you're not going to do it in this market. If you want to get ahead in this market on the air, the first thing you want to do is leave it."

He cited the example of Diane Barone, who was one of his pages for seven years, assigned to news. He explained, "She loved weather. She studied it in school; became a meteorologist. Did scripts, copy, research. She was very frustrated. She did a tape and went up to Washington and got a job up there for a year and a half, then she went to work for the 24-hour cable weather channel for another year and a half. She came back to L.A. and a weather spot opened up on Channel 9, and now she's back and she's very successful."

INTERNSHIPS

Nonpaying college internships do exist at the networks, and in L.A. at the networks' local stations, especially in their news departments. The advantage of the page programs over internships is that they do pay, they don't require that you get college credit, and, in some cases, like ABC, they aren't limited in time.

ENTRY LEVEL JOBS

There are also other entry level jobs at the networks. Elena Nachmanoff's job in New York is to hire for NBC News, for both on-air and off-air positions. When I spoke with her, she had just completed hiring for two network magazine shows. She says, "The entry level jobs are the production associates and program coordinators. I've hired people just getting out of school. I've hired people whose only experience is an internship. They come in and answer the phones. In their spare time, they come up with story ideas and are given research assignments. After a while, they move on to the next job, which may involve answering fewer phones. They may be promoted to an assistant producer position or a booker position. Their most important characteristic is that they have to have a 'can-do' attitude."

Lisa Simpson, director of business development and multimedia for NBC News in New York, handles all the department's business development work with the exception of the broadcast itself. She says that there are still entry level jobs in research and as receptionists, but that the network is looking for specialization, people who can deal with changing technology, specialized people with different skill sets, specifically in computers. She feels there are less-conventional ways to get into the business now.

MANAGEMENT PROGRAMS AND FELLOWSHIPS

Lisa was recruited through a special NBC Management Associates program. NBC is looking for future leaders, specifically ones with MBAs. While earning her degree in political science from USC, Lisa landed her first job working in the White House with Elizabeth Dole, lobbying for the new tax package in 1980.

After that, she worked on senatorial and presidential campaigns and ended up deputy regional director of the Reagan/Bush campaign.

Deciding she could perhaps do more in the private sector, she went back to school for a couple of years, earning an MBA from Harvard. She says, "Politics and media have a lot in common. You need a 'service gene,' a desire to make a difference, to help shape opinion."

Lynn Costa, director of organization management and resources planning for NBC in New York, says that NBC's Management Associates program is a one-year rotational program which gives associates a broad-brush exposure to marketing, finance and business development. Lisa proves the point that there are opportunities for new people in broadcasting if you've got special skills, entrepreneurial ideas, an eye on the future of the industry and a major-league work ethic.

In addition to its management associates page and internship programs, NBC also has a Minority Fellowship program, which provides scholarships to graduate students in journalism.

Like many of the other networks, including some of the cable networks and major production companies, NBC does *not* have an outside job line. Their policy is to promote from within as much as possible. So it all comes back to paying your dues, starting from the bottom, and networking like crazy.

Production Houses

Just as there's more to life than news, weather and sports, there's more to television than local stations, cable and networks. A tremendous amount of what we see on TV is produced by independent production companies and either supplied to broadcast and cable networks directly, or sold to one local station at a time in syndication.

You've got your prime-time series, TV movies, miniseries, game shows, news and entertainment magazines, sports specials, cooking shows, soap operas (although a couple are produced by the networks) and dozens of nationally syndicated talk shows.

There are literally thousands of outside producers. The giants—Buena Vista, Aaron Spelling, Carsey-Werner, Paramount, Lorimar, Ubu—are major operations with lots of production and hundreds, if not thousands, of employees. Some, like Harpo Productions, center around a particular talent. In addition to the daily "Oprah" show, Harpo (Oprah spelled backwards) has produced two movies-of-the-week, several after-school specials, Oprah's live interview with Michael Jackson and a couple of prime-time interview specials. Harpo employs 110 in their

Chicago facility. Other production companies are tiny operations trying to sell their first show.

Talk shows originate primarily in New York (Donahue, Geraldo, Sally Jessy Raphael, Maury Povich, Gordon Elliott, Regis & Kathie Lee), Chicago (Jenny Jones, Oprah) and L.A. (most of the rest). Talk show producers need staffers to handle their audiences. To give you an idea of the kinds of opportunities that exist in the production of a talk show, let's take the example of Oprah.

At Harpo Productions, entry level positions include pages, part-time positions often filled by local college students in the Chicago area; and staffers in the fan mail and research departments. Fan mail is a major resource for Harpo, as show ideas and potential guests are culled from the mail. It also takes major coordination to fill the audience with people who are interested in the subject matter and have stories of their own. Although some talk shows and production houses use interns, Harpo doesn't.

Breaking into Harpo is difficult because they don't post job openings for the general public and they only occasionally advertise their openings in the trades. Their jobs are posted internally as they like to promote from within. Many of those who land jobs at production companies attribute their success to luck and/or some kind of contact or recommendation from someone on the inside.

One exception to that rule is Andrea Wishom, audience supervisor at Harpo. After majoring in English at U.C. Berkeley, Andrea landed a coveted year-long minority trainee position, dividing her time among the three Chronicle Broadcasting stations, KRON in San Francisco, WOWT in Omaha and KAKE in Wichita, Kansas. When her year of writing, producing and editing ended, she worked hard to parlay that experience into her next job. One of the many resumes she sent out was to Harpo. Frustrated by a decided lack of job offers, Andrea applied and was verbally accepted into the Master's program at the prestigious Columbia School of Journalism at the University of Missouri.

Before the Master's program began, however, Harpo offered her a job as audience coordinator on the "Oprah" show. *U.S. News and World Report* actually reported on her turning down Columbia, a rare occurrence. But she admits that what she really wanted was a job. Within a year, Andrea was promoted to audience supervisor. She credits her landing a job at Harpo to the varied experience she gained working for three local affiliates as a Chronicle Broadcasting minority trainee.

Other production companies produce game shows, TV series and movies-of- the-week. Most are based in the Los Angeles area.

Suzy Polse Unger, vice president of development for Buena Vista Productions, the syndicated arm of Walt Disney Television, thinks that a great way to break into the business is by getting an entry level job with a production company, "holding cue cards, doing the day-to-day drudgery." She got her start in New York after interning with CNN.

She says, "There really are a lot of opportunities for people. With so many talk shows, people move up quickly. It's hard to find good people. The business is moving fast."

The best way to keep up with the ever-changing production industry is by reading the trades. *Variety* is one of the best resources for finding out who's doing what, and where.

TEN MAJOR PRODUCTION COMPANIES IN THE LOS ANGELES AREA

- *Buena Vista Television*
- *Stephen Bochco Productions*
- *Stephen J. Cannell Productions*
- *The Carsey-Werner Company*
- *Castlerock Entertainment*
- *Lorimar Television*
- *MCA-TV*
- *The Fred Silverman Company*
- *Aaron Spelling Productions, Inc.*
- *Warner Bros. International Television*

Doing Time

Once You're In

Congratulations! Your foot's in the door. You've got the job. You're thrilled. You're excited. You're gung-ho. And now your work begins.

Time to start laying the groundwork to get your next job. I'm assuming the job you landed isn't your ultimate goal. Even if you do stay in one position for a number of years, you'll need to work to keep it. You see, bosses come and go, and you'll need to prove yourself to each successive boss you're handed. In fact, you can be in the same position with the same boss and your job will change around you. With technological advances and ever-increasing competition for viewers and advertising dollars, you may have to run in place to keep up.

If you've landed an internship, temp job or minority trainee position, your time is even more precious and limited. You've got to learn all you can and make contacts and friends before your position expires.

LEARN WHO'S WHO

The first thing you want to do, after you've found the bathroom and the coffee machine, is learn the names of the other employees and their jobs/ranks/responsibilities ASAP. Make a list. Take it home and study it. You'll make a very good impression if you know who's who from the get-go. Conversely, you'll make a very bad impression when you're interning on the assignment desk and one of your own news producers calls in, and you're asking for the third time, "*Who* are you and who are you with?"

RUN

Joe Fragola of BayTV says, "You want to come in, you want to work hard. And it's really difficult. When I got into the business, I was a white middle-class male—and that's just a fact of life—and at the time, broadcasting had finally realized that there are people of color and women in the world. The competition for jobs grew tremendously. You had to do something that made you stand out. You had to be dependable. You had to have attitude, and you had to be able to run. So whenever I ran errands, I *ran* errands. I literally ran down the halls."

Running works. The director of the five o'clock news at KRON, Fred Bushardt, got his start some twenty years ago in the mailroom. He hit the ground running, and to this day his nickname is Fast Freddy.

SET GOALS AND FIND MENTORS

Once your foot is in, one of your first assignments is setting your next goal. Do you like the business? What areas are of greatest interest to you? What can you learn that will help you gain access to those areas? If you get a job as a news assistant or desk assistant and your goal is to move up in news, you want to improve your writing skills. Find a mentor or two and practice writing news stories, after hours if need be. Get someone good to critique your writing and work at it, all the while doing the job you've been hired to do to perfection.

Jim Gaughran, the Stanford English major bent on a journalism career mentioned earlier, spent 14 months in KRON's mailroom, during which time he got to know Darryl Compton, the associate news director who hired entry level newsroom staffers. He reminded Darryl regularly of his interest in news. Meanwhile, he made friends with the producers in the (now-defunct) children's programming department. They let him get involved in one of their specials, researching Native Americans and government funding. He also scored some points doing research on his own time for a public affairs program hosted by news reporters.

Jim's initiative impressed Darryl, who hired him as news assistant on the 3:00 to 11:00 p.m. shift. During the next year or so, Jim did his job, and talked a harried producer into allowing him to do a little newswriting "under the table." The 11 o'clock producer became his mentor. He recalls that his first story that made air was about a woman who had produced and was trying to market individually wrapped slices of peanut butter.

At the time, KRON had a one-year junior writer position, and after six months of "volunteer" writing, he landed that traineeship. Four months into that position, he was hired as a first-year writer. Four months later, he was promoted to second-year writer, and a year later, third-year writer. (Those are union designations indicating level of responsibility and pay.) Jim ended up head writer for five years, then, to expand his experience, he accepted the "Daybreak" newscast producer position. His shift was now midnight to 8:00 a.m. After six months of graveyard shift, he heard through a friend about a writer/editor job at "CBS Morning News." He landed the network job, and after 2 1/2 years was promoted to producer of "CBS Morning News." His hours were still graveyard, from 1:00 to 9:00 a.m.

Jim decided he preferred the West and the great outdoors and gave up the New York network position to come full circle, returning to KRON, where he started in 1981 in the mailroom. He's now producer of the 11:00 p.m. news and happy to be home.

VOLUNTEER ON THE INSIDE

Providing another example of volunteering from the inside, Shari Jackman, who worked as an intern in the sports department at KRON before landing a news assistant job, decided to volunteer her writing skills in the PR department. She didn't see much opportunity to advance from news assistant to newswriter within the station and had decided she was committed to staying in the Bay Area, where her family and fiancé lived. She said, "I knew I could write well, so I looked for other outlets in the station. One day, I went to lunch with Jodie Chase, manager of media relations, and she told me to come in and help. So since I worked from 3:30 to 11:30 p.m. in news, I would come in early in the morning and write press releases for Jodie. Just by volunteering, I demonstrated I had the determination to move on. Jodie always looked out for me because she knew I was willing to go the extra step. Although she realized I would be very helpful in her department, she also told me about other jobs outside the station."

I was heading up KRON's PR department at the time, and was well aware of Shari's dedication, intelligence and writing ability. We offered her a temporary publicist position when the station publicist went on maternity leave. We suspected the new mom wouldn't return, which was the case, and Shari got the job on a permanent basis.

Shari says, "Getting a mentor is very important. Find someone who is willing to show you the ropes and tell you when there are positions you'd be right for."

NETWORKING

Once you're in, you do get a chance to prove yourself, your capabilities, dependability and willingness to go the extra mile. While you're gaining your experience, pay attention to station gossip. You'll hear if the competition is doing something new and exciting. When someone new is hired by your station, there may be an opening at the station they just left. Keep your ear to the ground. Read the trades (listed in chapter 11 under Resources).

When talking to people in other markets, ask them about jobs in their area.

After landing the job as Maury Povich's assistant, Libby Moore started her own networking group of 30 personal assistants in the TV and entertainment business in New York. She explains, "We meet once a month for a potluck dinner at someone's apartment. It's a lot of work to organize, but everyone enjoys it and gets a lot out of it." An added benefit is that she's developed contacts who can occasionally help her with the show.

You have more clout in your networking efforts if you have a job. Janette Gitler, whom you read about in the chapter on internship success stories and is now in charge of local programming and community relations at KRON, provides one of the best examples of combining clout with gutsy networking.

She says, "When I was growing up, my dad watched the 'Tonight Show' every night, so I thought I wanted to be Freddy DeCordova, the producer. I'd see him with his headset on and his clipboard. He was the only behind-the-scenes person I knew. During my first producing job at KSTP, Freddy DeCordova was coming to the Twin Cities for a golf tournament. So I called the golf tournament people and said, 'This is Janette Gitler. I'm a producer of 'Twin Cities Today' and I understand that Freddy DeCordova is coming in for your golf tournament and I have to meet him, so is there any way that you can arrange a situation so I can meet him?' They said yes. They told me to meet them at the clubhouse on Sunday at 2 o'clock. So I showed up at the clubhouse at 2 o'clock and they introduced me to Mr. DeCordova, and I had such gall that I pulled out my card and said, 'It's a pleasure to meet you, Mr. DeCordova. You're socializing now and I don't want to take up your time, but I'm planning on being in Los Angeles next Monday and was wondering if there was any way we could get together to talk.' Well, I had absolutely no plans of going to Los Angeles, but he said, 'Sure,' gave me his secretary's name, told me to call her and set up an appointment. So the next day, I called and set up an appointment for the following Monday, then called the airlines and booked a flight to Los Angeles."

Once there, the clever Janette turned it into an informational interview, learned more about the network and the 'Tonight Show,' and made a valuable contact. It takes *chutzpah* and social skills to turn brief encounters into interviews and, possibly, jobs.

TAKE ON ADDITIONAL RESPONSIBILITY

A few months after Joe Fragola got his big break at CBS working on the Cronkite show in New York, he learned of an opening on the graveyard assignment desk for CBS News. He said, "It was just as a production assistant, loading the wire copy machines, sending telexes, calling and waking up crews all around the world, setting up the conference call for all the network executive producers to tell them what was going on that day before they came in at 6 o'clock in the morning." He took advantage of the opportunity to learn more and landed the job. Once established, he took on additional responsibility as overnight researcher, and when it was time to update his resume, he was able to list the position as overnight researcher/production assistant because he had refined and expanded the job. He says, "The overnight researcher is something I added on myself. I simply went up there on my own. Being aggressive like that is important. I got more experience. I think it paid off."

About news, Joe says, "Once you know the mechanics, you've got to develop instinct. What you have to do is know what you have to ask—'Why is this important?' 'What is this about?' 'Why should I care?'—the typical questions. And then sometimes you have to put things on because there's a need to know. There's a critical need to know. Also, you have to develop your instincts to answer the question, 'Is this person telling me the truth, is this person not telling me the truth?'"

FLATTERY WILL GET YOU SOMEWHERE

Speaking of the truth, you might want to bend the truth just a tad to get in good with your supervisors and others who can enhance your career opportunities.

Kissing up works. This fact was reported in Leah Garchik's "Personals" column in the *San Francisco Chronicle* on April 24, 1994. And I quote: "**How to Succeed in Business.** Psychologist Ron Deluga of Bryant College in Rhode Island has concluded, after studying the on-the-job behavior of 152 pairs of supervisors and subordinates, that kissing up works. According to Deluga, whose work is reported in the scholarly journal *Family Circle*, sympathizing, agreeing with, flattering and praising the boss gives an employee a 4 to 5 percent advantage over the drones who are relying on simple performance to get ahead. The most widespread way of kissing up, said the study, is agreeing with the boss." Garchik ended the item with "'Personals' hopes this item meets with the approval of her wise and witty supervisors."

All bosses are looking for employees who will make them look good. Most bosses have some insecurities of their own and are threatened by employees who are smarter than they, or want their jobs. Some, who have confidence and intelligence, *like* employees who have a lot of ideas of their own, and even appreciate someone with opposing views. This is rare.

One of the brightest, most creative people in the business I've met in 20 years is Janette Gitler. She's also one of the most outspoken. She gets away with it because she's damn good and, if not indispensable at one station, definitely an asset to the next. If you're as good as Janette, you can disregard the advice about kissing up.

Once you're in, no matter where you are in the pecking order, remember to give compliments when appropriate, give credit where credit is due, don't scream at people in front of their coworkers, take responsibility for your own mistakes, and generally don't step on others to climb the ladder of success. Don't forget the Golden Rule. There is a universal law of cause and effect, and what you put out will come back to haunt or reward you, as the case may be. You reap what you sow.

TEN WAYS TO GET AHEAD ONCE YOU'RE IN

- *Learn names, responsibilities, ranks of coworkers*
- *Learn your job quickly*
- *Do your job well*
- *Take on additional responsibility*
- *Compliment others for good work*
- *Take initiative*
- *Answer calls and correspondence promptly*
- *Be cheerful (don't complain)*
- *Don't make excuses or blame others*
- *Don't gossip*

Fast Forward to the Future

One of the joys of a television or cable career is that there's no such thing as boredom. The industry is moving faster than a speeding broadcast signal, or perhaps more appropriately, a digitized fiber-optic message. And clever people, who read the business pages, the TV columns and the trades, and pay attention to trends, can take advantage of the expanding opportunities.

The television industry was king of the communications jungle for its first 50 years. George Gilder[1] says, "[Television] contributed more to the U.S. standard of living than any other single invention. ... In terms of access to news and entertainment, television made the poorest of American families far richer than kings and tycoons of old."

According to Gilder, "The television age is giving way to the much richer, interactive technologies of the computer age. ... The new system will be the telecomputer, a personal computer adapted for video processing and connected by fiber-optic threads to other telecomputers all around the world. Using a two-way system of signals like telephones do, rather than broadcasting one-way like TV, the telecomputer will surpass the television in video communication just as the telephone surpassed the telegraph in verbal communication." Gilder predicts that the telecomputer's influence on American life and culture will be as great as the impact of television a couple of decades earlier. And rather than exalting mass culture, it will enhance individualism and promote creativity.

Shirley Davalos, who progressed from producing San Francisco TV talk shows to producing a national computer show for CKS Pictures, agrees. She says, "Television production as we know it is going the way of the buffalo. If you're going to succeed in the future, you must have an understanding of computers and multimedia, because the television age and the computer age are merging."

Keith Ferrell[2], vice president and editor of *Omni* magazine, wrote in the September 1994 issue of *Electronic Entertainment* magazine, "Our relationship to interactive platforms and programs is far more intimate than television can ever be. Every disc or cartridge is, in effect, a channel unto itself, with the advantage that its producers don't have to support a huge broadcast or cable distribution system. There is plenty of room for innovation and adventurous design, but we—the early audience—are the ones who have to make it clear that innovation and adventure are what we want."

Smart managers at television stations, networks and cable operations are adapting to the times. Lisa Simpson, director of business development and multimedia for NBC News in New York, is focusing on opportunities for the network news division in the future, looking into such areas as home video, interactive on-line and CD-ROM.

In terms of news and entertainment, past president of Fox News Van Gordon Sauter predicts that we'll be able to call up on our TVs the news and programming we wish to see when we wish to see it. We'll be able to call up "60 Minutes" any time of the week. "Network news will be gone. People will not wait. News will be local," Sauter suggests.

But Bill Groody says, "Whatever the distribution means, whether it's cable, free-radiating television, or something shot down from a satellite, there will always be a need for news and information programming. It's going to be around a long time."

Broadcasting will continue to give way to narrowcasting, where the consumer will be in charge. Opportunities will continue to expand as cable and telecommunications continue to merge.

Whatever your vision of the future, whatever the impact of fiber-optics and the new technologies in telecommunications and computers, there is no question that the changes will be enormous and affect all aspects of our way of life. Our options in terms of how we use our time, both business and pleasure, will be just this side of infinite. We consumers in the megachannel universe of the future will be faced with megadecisions about which resources we want to access at any given moment.

And those of you who decide to work in the communications industries of the future will be creating the new world that will benefit us all.

NOTES

[1] George Gilder, *Life After Television* (Knoxville, TN: Whittle Direct Books, 1990).

[2] Keith Ferrell, *Electronic Entertainment*, September 1994, p. 32.

TEN RELATED INDUSTRIES/BUSINESSES/CAREERS

- *Advertising*
- *Media sales (rep firms)*
- *Multimedia*
- *Radio*
- *Film*
- *Corporate in-house video production*
- *Theater*
- *Public relations*
- *Special event production*
- *Talent agencies*

A Final Word

I would love to have feedback from you on your search, whether or not you end up in television. Please let me know what worked or didn't work for you, and perhaps I can pass on your advice to the next generation of broadcasters in a future edition of this book.

Here are a variety of ways to contact me.

Linda Farris
Buy the Book Enterprises
182 Canyon Road
Fairfax, CA 94930

Fax: (415) 454-4829

E-mail: guessgenes@aol.com

Good luck, and write if you get work!

Appendix

A. U.S. MARKETS RANKED BY SIZE

Source: *Broadcasting & Cable Yearbook* 1994

1. New York, NY
2. Los Angeles, CA
3. Chicago, IL
4. Philadelphia, PA
5. San Francisco-Oakland-San Jose, CA
6. Boston, MA
7. Washington, DC
8. Dallas/Ft. Worth, TX
9. Detroit, MI
10. Atlanta, GA
11. Houston, TX
12. Cleveland, OH
13. Minneapolis-St. Paul, MN
14. Seattle-Tacoma, WA
15. Miami-Ft. Lauderdale, FL
16. Tampa-St. Petersburg, FL
17. Pittsburgh, PA
18. St. Louis, MO
19. Sacramento-Stockton, CA
20. Phoenix, AZ
21. Denver, CO
22. Baltimore, MD
23. Orlando-Daytona Beach-Melbourne, FL
24. Hartford-New Haven, CT
25. San Diego, CA
26. Portland, OR
27. Indianapolis, IN
28. Kansas City, MO
29. Milwaukee, WI
30. Charlotte, NC
31. Cincinnati, OH
32. Raleigh-Durham, NC
33. Nashville, TN
34. Columbus, OH
35. Greenville, SC-Spartanburg, SC-Asheville, NC
36. San Antonio-Victoria, TX
37. Grand Rapids-Kalamazoo-Battle Creek, MI
38. Buffalo, NY
39. Norfolk-Portsmouth-Newport News-Hampton, VA
40. New Orleans, LA
41. Salt Lake City, UT

42. Memphis, TN
43. Providence, RI-New Bedford, CT
44. Harrisburg-York-Lancaster-Lebanon, PA
45. Oklahoma City, OK (tie)
45. West Palm Beach-Ft. Pierce-Vero Beach, FL (tie)
47. Louisville, KY
48. Greensboro-Winston Salem-High Point, NC
49. Birmingham, AL
50. Wilkes Barre-Scranton, PA
51. Albuquerque, NM
52. Albany-Schenectady-Troy, NY
53. Dayton, OH
54. Jacksonville, FL
55. Charleston-Huntington, WV
56. Fresno-Visalia, CA
57. Flint-Saginaw-Bay City, MI
58. Little Rock, AR
59. Tulsa, OK
60. Richmond, VA
61. Wichita-Hutchinson, KS
62. Knoxville, TN
63. Mobile, AL-Pensacola, FL
64. Toledo, OH
65. Green Bay-Appleton, WI
66. Austin, TX
67. Roanoke-Lynchburg, VA
68. Syracuse, NY
69. Rochester, NY
70. Des Moines, IA
71. Shreveport, LA-Texarkana, TX
72. Lexington, KY
73. Omaha, NB
74. Springfield-Decatur-Champaign, IL
75. Portland-Poland Spring, ME
76. Paducah, KY-Cape Girardeau, MO-Harrisburg-Marion, IL
77. Las Vegas, NV
78. Springfield, MO
79. Tucson, AZ
80. Spokane, WA
81. Huntsville-Decatur-Florence, AL
82. Cedar Rapids-Waterloo-Dubuque, IA
83. South Bend-Elkhart, IN
84. Davenport-Rock Island-Moline-Quad City, IA
85. Chattanooga, TN
86. Columbia, SC
87. Jackson, MS
88. Ft. Myers-Naples, FL
89. Johnstown-Altoona, PA
90. Bristol, VA-Kingsport-Johnson City, TN-Tri-City
91. Madison, WI
92. Youngstown, OH
93. Burlington, VT-Plattsburgh, NY

94. Evansville, IN
95. Baton Rouge, LA
96. Waco-Temple-Bryan, TX
97. Springfield, MA
98. Colorado Springs-Pueblo, CO
99. Lincoln-Hastings-Kearny, NB
100. El Paso, TX
101. Ft. Wayne, IN
102. Savannah, GA
103. Greenville-New Bern-Washington, NC
104. Lansing, MI
105. Charleston, SC
106. Peoria-Bloomington, IL
107. Sioux Falls-Mitchell, SD
108. Fargo, ND
109. Santa Barbara-Santa Maria-San Luis Obispo, CA
110. Montgomery-Selma, AL
111. Augusta, GA
112. Tyler-Longview-Jacksonville, TX
113. Salinas-Monterey, CA
114. McAllen-Brownsville, TX
115. Tallahassee-Thomasville, FL
116. Reno, NV
117. Ft. Smith, AR
118. Lafayette, LA
119. Traverse City-Cadillac, MI
120. Macon, GA
121. Columbus, GA
122. Columbus-Tupelo, MS
123. Corpus Christi, TX
124. Eugene, OR
125. Duluth-Superior, MN
126. La Crosse-Eau Claire, WI
127. Yakima-Pasco-Richland - Kennewick, WA
128. Amarillo, TX
129. Monroe, LA-El Dorado, AR
130. Chico-Redding, CA
131. Bakersfield, CA
132. Wausau-Rhinelander, WI
133. Boise, ID
134. Binghamton, NY
135. Wichita Falls, TX-Lawton, OK
136. Rockford, IL
137. Topeka, KS
138. Terre Haute, IN
139. Florence-Myrtle Beach, SC
140. Beaumont-Port Arthur, TX
141. Sioux City, IA
142. Wheeling, WV-Steubenville, OH
143. Erie, PA
144. Wilmington, NC
145. Medford, OR
146. Joplin, MO-Pittsburg, KS
147. Rochester, MN-Mason City, IA-Austin, MN
148. Bluefield-Beckley-Oak Hill, WV (tie)
148. Lubbock, TX (tie)

150. Minot-Bismarck-Dickinson, ND-Glendive, MT
151. Columbia-Jefferson City, MO
152. Odessa-Midland, TX
153. Sarasota, FL
154. Albany, GA
155. Bangor, ME
156. Abilene-Sweetwater, TX
157. Idaho Falls-Pocatello, ID
158. Biloxi-Gulfport-Pascagoula, MS
159. Quincy-Hannibal, MO
160. Utica, NY
161. Clarksburg-Weston, WV
162. Salisbury, MD
163. Panama City, FL
164. Gainesville, FL
165. Laurel-Hattiesburg, MS (tie)
165. Palm Springs, CA (tie)
167. Dothan, AL
168. Watertown-Carthage, NY
169. Rapid City, SD
170. Elmira, NY
171. Alexandria, LA
172. Harrisonburg, VA
173. Billings-Hardin, MT
174. Jonesboro, AR
175. Lake Charles, LA
176. Missoula, MT
177. Ardmore-Ada, OK
178. Greenwood-Greenville, MS
179. El Centro, CA-Yuma, AZ
180. Meridian, MS
181. Jackson, TN
182. Great Falls, MT
183. Grand Junction-Durango, CO
184. Parkersburg, WV
185. Tuscaloosa, AL
186. Eureka, CA (tie)
186. Marquette, MI (tie)
188. San Angelo, TX
189. Butte, MT
190. Lafayette, IN
191. Bowling Green, KY
192. Hagerstown, MD
193. St. Joseph, MO
194. Anniston, AL
195. Cheyenne-Scottsbluff, WY (Sterling)
196. Charlottesville, VA
197. Casper-Riverton, WY
198. Laredo, TX (tie)
198. Lima, OH (tie)
200. Ottumwa-Kirksville, IA
201. Twin Falls, ID
202. Zanesville, OH
203. Presque Isle, ME
204. Bend, OR
205. Mankato, MN
206. Flagstaff, AZ
207. Helena, MT
208. North Platte, NB
209. Alpena, MI

B. ACRONYMS IN TELEVISION

AAJA	Asian American Journalists Association
AE	Account executive
AWRT	American Women in Radio and Television
CNN	Cable News Network
FCC	Federal Communications Commission
GRP	Gross rating point
HUT	Homes using television
IBEW	International Brotherhood of Electrical Workers
LO	Local origination
MIS	Management information systems
MSO	Multiple system operator
NAB	National Association of Broadcasters
NABET	National Association of Broadcast Employees and Technicians
IATSE	International Alliance of Theatrical Stage Employees and Motion Picture Machine Operators
NABJ	National Association of Black Journalists
NAHJ	National Association of Hispanic Journalists
NAMIC	National Association of Minorities in Cable
NATAS	National Academy of Television Arts & Sciences
NATPE	National Association of Television Program Executives
NCTA	National Cable Television Association Inc.
O&Os	Stations owned and operated by a network
PBS	Public Broadcasting System
PSA	Public service announcement
RTNDA	Radio-Television News Directors Association
SBE	Society of Broadcast Engineers
SOT	Sound on tape
SPJ	Society of Professional Journalists
TBS	Turner Broadcasting System, Inc.
TD	Technical director or switcher
VJ	Video journalist
VO	Voiceover
WICI	Women in Communications, Inc.

Index

T

U

V

W

Z

About the Author

Linda Guess Farris was born in Hollywood and raised on a cattle ranch in Texas. In 1969, she received a B.A. in sociology from Smith College.

Settling in San Francisco later that year, Farris joined Nob Hill Travel in the Fairmont Hotel as a travel agent and tour escort to Hawaii and Asia. From 1970 to 1974, Farris worked as a secretary in the promotion and creative departments of Clinton E. Frank Advertising. Farris broke into TV in 1974 as a secretary/listings editor/publicist at KGO-TV, San Francisco's ABC station. A couple of years later, she was promoted to manager of press information.

In 1979, Farris moved to KRON-TV, San Francisco's NBC affiliate, as director of public relations. She was promoted to director of marketing services ten years later.

Farris won the national Crane Communication Award for Excellence in Television & Radio Promotion & Marketing in 1986. Also that year, she was honored by the San Francisco chapter of American Women in Radio & Television with an Achievement Award for her contribution to broadcasting.

During her twenty-year career in the industry, Farris regularly conducted career seminars and spoke to groups about breaking into television.

Since 1993, Farris has been writing and publishing books and freelancing in Bay Area television.